CULTIVATION AND AGRONOMY OF MEDICINAL PLANTS

CULTIVATION AND AGRONOMY OF MEDICINAL PLANTS

Dr. P. K. Bhagat

SWASTIK

SWASTIK PUBLICATIONS

DELHI–110094 (INDIA)

SWASTIK PUBLICATIONS
31 Gali No. 1, A Block, Pocket 5, CRP Water Tank,
Sonia Vihar, Delhi-110094
Cell: 9968482939, 9899462604
e-mail: swastik_books@yahoo.com

Cultivation and Agronomy of Medicinal Plants

First Published 2012

ISBN:- 978-93-81084-27-4

PRINTED IN INDIA

Published by Indra Mani Pandey for Swastik Publications, Delhi and Printed at Asian Offset Press, Delhi-53

Preface

Cultivation of Medicinal Plants is gaining ground because of the sky rocketing prices of allopathic medicines which also have side effects. Cultivation of medicinal plants is economically very attractive. In recent years, there has been a tremendous growth of interest in plant-based drugs, pharmaceuticals, perfumery products, cosmetics and aroma compounds used in food flavours and fragrances and natural colours in the world. There is a definite trend to adopt plant-based products due to the cumulative derogatory effects resulting from the use of antibiotics and synthetics; and except for a few cultivated crops, the availability of plant-based materials is mainly from the natural sources like forests and waste-lands. But, due to the ever increasing utilization of land for food crops, devastation of forests and the concurrent indiscriminate exploitation of these crops, their availability from natural sources has declined. On the other hand, the demand for internal use and for export has been increasing, necessitating the production of these crops on a large scale. There is a need to introduce these crops into the cropping systems of the country which, besides meeting the demands of the industry, will also help to maintain the standards on quality, potency and chemical composition.

Since these crops are new and uncommon to the growers and they often lack knowledge of their scientific cultivation, there is hesitation even among the interested growers, to take up their cultivation. An attempt, therefore, has been made in this book to provide all possible pooled information including the research findings that have generated through research by the Division of Horticulture, University of Agricultural Sciences, Indian Institute of Horticultural Research, Central Institute of Medicinal and Aromatic Crops, National Botanical Research Institute, Regional Research Laboratories, ICAR and others.

The author also takes this opportunity to thank the various past and present post-graduate students in the field of medicinal and aromatic crops who have been instrumental in providing research information and also in collecting the information from various sources. Hope this attempt will bear fruit for the farmers and persons devoted in this field.

Author

CONTENTS

Introduction

Medicinal and aromatic plants constitute a major segment of the flora, which provides raw materials for use in the pharmaceuticals, cosmetics, and drug industries. The indigenous systems of medicines, developed in India for centuries, make use of many medicinal herbs. These systems include Ayurveda, Siddha, Unani, and many other indigenous practices. More than 9,000 native plants have been established and recorded for their curative properties and about 1500 species are known for their aroma and flavour. In one of the studies by the World Health Organisation, it is estimated that 80 per cent of the population of developing countries relies on traditional plant based medicines for their health requirements (WHO, 1991).

Even in many of the modern medicines, the basic composition is derived from medicinal plants and these have become acceptable medicines for many reasons which include easy availability, least side effects, low prices, environmental friendliness and lasting curative property. India and China are the two major producing countries, having 40 per cent of the global biodiversity and availability of rare species. These are well known as the home of medicinal and aromatic crops that constitute a segment of the flora, and provide raw materials to the pharmaceutical, cosmetic, fragrance, flavour etc. industries.

The Ministry of Environment and Forest, Government of India has identified and documented over 9,500 species of medicinal plants that are significant for the pharmaceutical industry. Of these, 2,0 00 to 2,300 species are used in traditional medicines while at least 150 species are used commercially on a large scale (EXIM Bank, 1997). The fact that derivatives of medicinal and aromatic plants are non - narcotic have no ill effects and constitute the natural base for treatments, have resulted in an increase in demand for these plants in developing and non developed countries. Due to this rising international demand, many important medicinal plant species are becoming scarce and some are facing the prospect of extinction.

Therefore, it is important to conserve the extensively traded medicinal plants in its natural environment or cultivate it in favourable environments. As compared to the traditional crops, the cultivation of medicinal crops has many advantages. These include:

- Medicinal crops provide better returns than traditional crops;
- Have very high domestic and export demand;
- Fetch better prices in the market with lower price fluctuations;
- Could be stored for a long time, and sold at a time when better prices prevail in the market;
- These are largely drought tolerant, and not easily grazed by animals;
- Have low incidence of pest attacks and diseases;
- Require minimum resources, therefore the cost of cultivation is lower compared to the traditional crops;
- Could be raised as inter-crops, along with traditional crops, and also on degraded lands.

Given these advantages the cultivation of medicinal crops has been picking up in some regions. However, the spread is not so large to meet the demand of the industry. Owing to short supply, prices of some medicinal crops have increased substantially in the recent past. What is also of great concern is that overexploitation of some of the species has threatened their extinction.

Importance of the Medicinal Crops

The Indian farmer is presently in quite a difficult economic tangle and majority of them are looking at options other than farming (NSSO, 59 th round). Farm household today is constrained by the shrinking net income flow due to the increasing cost of cultivation and uncertainty in prices. Largely, for long marginal and small farmers have not diversified from traditional crops. Production costs of traditional crops are now increasing due to the larger share of cash inputs and the market system is unsupportiv e. These factors have forced the farmers to shift towards other crops that have better potential returns compared to the traditional crops. In this backdrop, the cultivation of medicinal crops is less risky in terms of the incidence of pest attacks, diseases and price fluctuations, and has potential returns.

And this significance is further strengthened by the fact that these crops could be grown in degraded and marginal soils, or raised as inter crops in plantation crops like arecanut, coconut etc. with less difficulty. The trade demand for these crops is also increasing with the increased interest in western consumers towards eastern medicinal systems. The estimated area under the medicinal crops in India is in the neighbourhood of two lakh hectares. Nearly 75 per cent of the plant material used in indigenous medicines is collected from forests and wild habitats (GoI, 2000). People hardly have any knowledge about the plants, collect the material and pay little attention to selective harvest or natural harvesting of plants.

Traditionally, India has been considered a treasure house of valuable medicinal and aromatic plant species. The Indian System of Medicine uses over 1,100 medicinal plants and most of them are collected from forests regularly, and over 60 species among them are particularly in demands (GoI, 2000). On account of the fact that derivatives of medicinal and aromatic plants have no side effects and deal curatively, the demand for these plants is on the increase in both developing and developed countries. As a result, the trade of medicinal plants is increasing fast.

From the trade data available, it is clear that the global market for medicinal plants has always been large and has been increasing in the recent past. In the report commissioned by the World Wide Fund for Nature, it is pointed out that, the total import in 1980 of "vegetable materials used in pharmacy" by the European Economic Community was 80,738 tons (Lewington 1993). India was the largest supplier with 10.05 tons of plants and 14 tons of vegetable alkaloid and their derivatives. India, Brazil and China are the largest exporters of medicinal plants.

Trade of medicinal plants from India is estimated to be worth Rs.550 crore and likely to reach Rs.1,000 crores. Cosmetics and aromatherapy products are two important areas where Indian medicinal plants and their extracts like essential oils can contribute globally. Medicinal and aromatic plants have a high market potential with the world demand for herbal products growing at the rate of seven per cent per annum. There is now wide recognition of the contributions that medicinal and aromatic plants partake in the global economy and human welfare (WHO, 1978). But one of the greatest difficulties in assessing their importance as resources, either locally or globally, is the shortage of dependable information about the species being used, their availability and distribution, how these are collected or harvested, where the species are cultivated, the quantities involved and trade statistics. Con fidentiality still surrounds the industry as well as the State run systems. Most of the evidence is anecdotal, although there has been a concentrated effort in recent years to gather information on these aspects not many documents are av ailable. National and regional assessments have been published for several countries; also various global reports have been prepared such as Husain (1996) on international trade, marketing and consumption of essential oils, and McAlpine et al. (1997) on future world trends in supply, utilisation and marketing (Orient Longman).

IMPORTANCE, PRESENT STATUS AND FUTURE PROSPECTS OF MEDICINAL CROPS

Plants have been one of the important source of medicines even since the dawn of human civilization. In spite of tremendous developments in the field of allopathy during the 20th century, plants still remain one of the major source of drugs in modern as well as traditional systems of medicine throughout the world. Approximately one-third of all pharmaceuticals are of plant origin, wherein fungi and bacteria are also included. Over 60% of all pharmaceuticals are plant-based.

Chemically, depending on their active principles, plants may have alkaloids, glycosides, steroids or other groups of compounds which may have marked pharmaceutical actions as anticancerous, anti-malarial, antihelminthic or antidysentric, etc. Many of the essential oils, dyes, lattices and even tannins and vegetable oils are also widely used as medicines. The many substances that go into making up medicines are frequently products of living cells, although seemingly 'waste' or intermediate, metabolic compounds and not an integral part of the protoplasm and may have no obvious utility to the plants.

Out of nearly 3 00 000 species of higher plants available, only a small proportion have been investigated for medicinal properties, and a still smaller number yield well-defined drugs. The same is the case with lower plants and with plants of the sea. Thus, the knowledge of plant constitutents gained so far is still meagre, considering the huge number of species available in the world. &proximately, only 10% of the organic constitutents of lants are reported to be known and the remaining 90% are yet to be explored.

A very small proportion of Indian medicinal plants are lower plants like lichens, ferns, algae, etc. The majority of medicinal plants are higher plants. The major families in which medicinal plants occur are *Fabaceae, Euphorbioceoe, Asteraceae, Pooceae, Rubiaceae, Cucur-bitaceae, Apiaceae, Convolvulaceae. Malvaceae* and *Solanaceae.*

Drugs are derived from trees, shrubs and herbs and even from primitive kinds of plants which do not fall into the above categories. They are made from fruits (Senna, *Solanum viarum,* Datura, etc.), flowers *(Butea monosperma, Bauhinia verigata),* leaves (Senna, Datura, Periwinkle, *Tyloph* ora , etc.), stems (Liquorice, Ginger, Dioscorea, Costus,Garlic), roots (Rauvolfia, Periwinkle, Ginseng, etc.), seeds (Isabgol, *Abrus, Nux vomica)* and even bark (Cinchona).

History

Plants have been associated with the health of mankind from time immemorial. In the past, sickness was viewed as a punishment from the gods and, hence, was treated with prayers and rituals that included what may have been considered 'magic potions' prepared from local herbs.

Archaeological discoveries of 60,000-year-old Neanderthal burial grounds in Iraq, point to the use of several plants like marsh mallow, yarrow and groundsel that still figure in folk medicine. Mexican Indians are reported to have used peyote cactus for its hallucinogenic and, also possibly, its healing properties for thousands of years. This plant is now known to have antibiotic properties as well.

The Sumerians of the Tigris and Euphrates (present day Iraq) around 4000 BC from their cuneiform writing on clay tablets are reported to have used opium, liquorice, thyme, mustard and the chemical element sulphur as medicine. The Babylonians who apparently followed the Sumerians in this field added senna, coriander, saffron, cinnamon and garlic among the other herbs in their formulations.

The Contribution of the Egyptians

Next, out of the great civilization of Egypt, came Imhotep, a skilled physician who later became the Egyptian god of healing. Ancient Egypt also gave the world one of its first medical texts, the *Ebers payrus* named after the German Egyptologist George Ebers. This payprus is believed to have been written in the 16th century BC. It contains about 800 recipes and refers to over 700 drugs including aloe, wormwood, peppermint, henbane, myrrh, hemp, dogbane, castor oil and mandragora. With such ingredients, they prepared wines and infusions, as well as pills, salves and poultices. The *Ebers Papyrus* mentions a recipe that suggests that the Egyptians had a treatment for diabetes. It also advises the placing of mud or mouldy bread over sores to keep them from becoming infected. Not until a millennium later was it discovered that mud and mould often contain certain micro-organisms like filamentous bacteria and filamentous fungi that produce a class of antibiotic wonder drugs.

The Contribution of the Chinese

Ancient Egypt was not alone in demonstrating the healing powers of plants. The Chinese pharmacopoeia, *Pen-T Sao Kens Mu,* is the old-est record of herbal medicine that we know of, and is reported to have been written by the Chinese Emperor Shen Nung (3737-2697 BC). This work describes the use of chaulmoogra oil from trees of the *Hydrocarpus* genus to treat leprosy. Among the many plants listed by them, opium poppy as well as rhubarb and aconite are included. This ancient record also includes the use of a desert shrub called Ephendra or Mahuang to improve circulation, reduce fevers, help urinary functions, supress coughs and relieve the lungs of bronchial disorders. Its active ingredient was nearly lost to modern medical science until its rediscovery earlier in this century. We now know it as Ephedrine, the key ingredient in modern pharmaceuticals, which is used to relieve breathing difficulties and other symptoms of asthma, hay-fever and common cold.

The Contribution of the Greeks

Ancient Greece had a god of healing named Aesculapius. His sign was a snake curled around a staff—the caduceus, which is still a symbol of medicine today. Medicine was practiced by priests called "Aesculapiods" and the treatment used to be a religious ritual full of incantations and mystery, carried out over several days of fasting and bathing.

In 400 BC, a Greek named Hippocrates separated the healing profession from the realm of mysticism and religion. He asserted that medicine was a science and an art. In his teachings, he laid great emphasis on diet, lifestyle, exercise, sunshine and water. His famous oath suggests strong humanitarian concerns among the Greek physicians of the era. For his contribution, he is called the 'Father of Modern Medicine'. His writings name about 300-400 healing plants.

After Hippocrates came Aristotle, whose far-ranging scientific work included an effort to catalogue the properties of various medicinal herbs. Aristotle's pupil Theo hrast s 300 BC a botanist whose treatise *Inquiry Into Plants* influenced both botany and medicine for centuries, is called the 'Father of Pharmacognosy'. He described the characteristics of herbs like cassia, mentha, belladonna, squill and several others. He was also able to change the properties of some of these plants through cultivation (mutant varieties).

In the 1st century AD, Greece produced the forerunner of all modern pharmacopoeias and the most authoritative text on botanical medicine for over a thousand years, *De Materia Medica.* In this book, Dioscoroides (64-120 AD) discussed in detail the identification, collection, adulteration and therapeutic uses of several thousands of plants.

Pliny's *Natural History,* published in the 1st century AD, was the compilation of thousands of Greek and Roman treatises. Much of what Pliny wrote down eventually passed into the folklore of Europe and the New World. The *Natural History* stated that plants exist to meet man's needs and plants not clearly useful for man's food, clothing and shelter may possess medicinal properties.

Out of the same period came yet another great figure of medicine, Galen, a Greek physician who practised in Rome during the 2nd century AD. Galen's fame is primarily associated with the fauna, not the flora, of his time and place. He revolutionized medicine by performing animal experiments, from which he developed the first medical theory based on scientific investigations. Although many of Galen's theories were proved wrong, because he assumed animal studies would directly apply to human beings, his place as the founder of 'Experimental Medicine' is unchallenged. He is credited with the divergent theories of Allopathic (cure a disease by its opposite) and Homoeopathic (like cure like disease) medicine, both of which have grown out of Galen's doctrines.

The Contribution of the Church

In Europe, from about 500 to 1500 AD, was a period in which almost all medicinal knowledge flourished. Herbalism continued largely undisturbed, although with a religious orientation. It was during this period that many plants acquired names linked to Jesus, the Virgin Mary, saints and martyrs. In the latter part of the medieval period, the church actively encouraged two major advances in medicine. One was the hospital, a system of caring for the sick without charge and their second great contribution to medicine was the establishment of the first 'University of Medical Schools'.

The Contribution of the Arabs

The Arabs translated the original works of the Greeks into their own language and made refinements based upon their own experiences. They added plants such as camphor, saffron and spinach to the classical pharmacopoeia.

Rhazes (Ar Razi), a physician born in Persia in the late 9th century, wrote a famous treatise correctly describing smallpox and measles for the first time. And about a hundred years later, Avicenna (Ibne sina), the prince of physicians was born. His *Canon of Medicine* contains many references to the teachings of Galen and Anstotle. The canon was used as a textbook in medical schools through-out Europe well into the 17th century and is still studied in the east. In it, Avicenna described meningitis, tetanus and many other diseases.

The Indian Contribution

The Aryans of the Indus Valley wrote three treatises, viz. the *Rigveda.* (2000 BC), *Atharvaveda* (2000-1000 BC) and *Aurveda* 100-600 BC which mention several medicinal-Plants and their uses, including the hallucinogenic mushrooms *Amanita muscaria* and *Rauvolfia serpentina* used to treat snakebite, epilepsy, mental disorders and other illnesses. This plant is the source of reserpine, a tranquillizer and hypertensive agent widely used in modern pharmacy. The *Charaka Samhita,* an encyclopaedia of Indian medicine! published at Varanasi between 1000 BC and 100 AD, is a comprehensive record of medicinal plants and their uses.

However, after 1200 AD, there was a period of decline in the investigation of local medicinal plants due to the invasion of the Arabs and others, and Arabic medicine which was also based mainly on plants attained more prominence. During this period many important medicinal plants like senna, isabgol, roselle, chirata, opium poppy, etc. were introduced into India by them.

From 1800 AD onwards, the European invasion resulted in the introduction of the Western allopathic system in India and this also included the use of the number of medicinal plants of proven therapeutic value.

The Development of Pharmacopoeia and Chemical Investigations

As regards a pharmacopoeia which lists the use of medicinal plants, the first official record, *Nuonova Receptro,* was published in france in 1408 AD Later, after about 50 years, pharmacopoeias in other countries made their appearance.

The chemical investigation of plants on scientific lines started from 1800 AD onwards and the important discoveries in this regard were those of F W Sertuner (1806), a German pharmacist, who discovered morphine from opium from the plant *Papaver somniferaum,* and the French chemists, P J Pellatier and J B Caventou (1820), who discovered quinine from the cinchona bark which was universally recognized as a cure for malaria. They are considered to have laid the foundation for the school of Plant Chemistry, which resulted in the discovery of many active principles of plants useful in the alleviation of human diseases.

The Importance of Plants

Plants have been used by man from prehistoric times for relieving suffering and curing ailments. Primitive people, when injured in battle or when they had a fall or cut, instinctively resorted to materials available at hand for staunching the flow of blood or the relieving of pain and, by trial and error, they learnt that certain plants were more effective than others for these purposes. Man has also gained such knowledge from his observation of birds and animals which use plants for curing their ailments. Even today, we find that the domestic dog and cat, when they suffer from indigestion or other ailments, run to the field, chew some grasses or herbs and get cured. The folk medicines of almost all the countries of the world abound in medicinal plants and tribal people wherever they exist, rely chiefly on herbal medicine, even today.

Today, chemical and pharmaceutical investigations have added a great deal of status to the use of medicinal plants by revealing the presence of the active principles and their actions on human and animal systems. Investigations in the field of Pharmacognosy and Pharmacology have supplied valuable information on medicinal plants with regard to their availability, botanical properties, methods of cultivation, collection, storage, commerce and therapeutic uses. All these have contributed towards their acceptance in modern medicine and their inclusion in the pharmacopoeias of civilized nations.

The indigenous systems of medicine practiced in India are based mainly on the use of plants. *Charaka Samhlta* (1000 Be-100 AD) records the use of 2000 vegetable remedies. Ancient medicine was not solely based on empiricism and this is evident from the fact that some medicinal plants which were used in ancient times still have their place in modern therapy. Thus, for example, 'Ephedra' a plant used in China 4 000 years ago is still mentioned in modern pharmacopoeias as the source of an important drug, ephedrine. The plant Sarpagandha *(Rauvolfia serpentina)* which was well known in India as a remedy for insanity has now shown that one of its constituents, reserpine, is a wonder drug today for curing mental ailments. Quinine, another important antimalarial drug of modern medicine, was obtained from the cinchona tree.

The knowledge about the use of medicinal plants has been accrued through centuries and such plants are still valued even today, although synthetics, antibiotics, etc. have attained greater prominence in modern medicine. It is, however, a fact that these synthetics and antibiotics although they often show miraculous and often instantaneous results, prove harmful in the long run and that is why many synthetics and antibiotics have now gone out of use or have been specified to be prescribed strictly under medical supervision. In the case of most medicinal plants, however, no such cumulative derogatory effect has been recorded and that is why many of the medicines obtained from plants are still widely used today.

It is also true that lately, due to the rapid progress and spread of modern medicine, the popularity of herbal medicines has been gradually declining. But in a country where 80% of our population resides in small and remote villages, and most of them are economically backward, complicated prescriptions which are usually expensive and difficult to procure cannot be a thing of the masses.

Besides the above, the following are some of the reasons that make the cultivation of medicinal plants very important.

1. In nature there remains a wide variation among the plants with regard to their active principles. As only the best among them are used for cultivation, it enables us to obtain raw material of homogenous quality and high potency.
2. It is easy to grow and fulfill the commitment of large-scale supply through cultivated sources rather than from natural sources which mainly depend on nature for their regeneration and availability.
3. The growing pressure of population/urbanization and the development of roads to remote areas has resulted in deforestation and the loss of natural plant resources.
4. In many cases, as the important plant parts used are roots or the entire plant, this results in plant collectors engaging in destructive collection/extractive methods, resulting in many plants becoming extinct or being listed as threatened.
5. Despite the fact that our forests are a major resource-base for medicinal plants as many of them appear in the wild, the importance of this has not been totally appreciated by the concerned government departments. So much so, that many forest action plans do not clearly spell out any long-term strategy for the conservation of biodiversity and support to the communities which are solely dependent on forests for their livelihood.

6. The unauthorized collection of minor forest produce by persons who are led by the burgeoning demand for raw medicinal plant parts has led to the deprivation of the rights and opportunities of the forest-dwelling communities.

Present Status

(a) Status *in* India

The age-old Indian systems of medicine, have been neglected mainly because of the rapid expansion of the allopathic system of medical treatment. This is despite the fact that our country has a long history of local health traditions, which are backed by thou-sands of scriptures left behind by practitioners of these systems of medicine. One of the earliest treatises of Indian medicine, the *Charaka Samhita* (1000 BC), mentions the use of 2000 vegetable herbs for medicinal use. Over 7000 different species of plants found in different ecosystems are said to be used for medicinal purposes in our country. The traditional Indian system of medicine can be broadly classified into the empirical forms of folk medicines which are village-based, region-specific, indigenous herb-based, local resource -based and, in many cases, community-specific.

The other system called the *Shastriya* stream which includes the *Ayurveda, Siddha* and *Unani* systems of medicine is said to be more complicated and elaborate with theoretical and research findings. It is also said to be documented in thousands of regional manuscripts. About 100,000 medical manuscripts are said to be available in oriental libraries and private collections in India and abroad, but only a few hundreds of these books are made available to students and teachers of Indian medicine.

Presently, the Indian system of medicine uses over 1100 medicinal plants and most of them are collected from the wild regularly, of which over five dozen species are said to be in great demand. The tribal belt of India is rich in these plants and local tribes mainly depend, for their livelihood, on their collection and trade. However, the supply of raw material for the industry is procured through minor forest produce contractors/dealers. They, in fact, determine the prices of the material depending on the season, demand, availability, etc. Moreover, in many cases, there are no standards fixed for the raw materials and, physical appearance verification generally becomes impossible because of the large quantities of material involved. Since these plants are collected from the wild in large quantities, adulterants are alleged to be widely used.

The collectors/tribals settled in forests who do the collection are generally paid only subsistence allowances (less than the prescribed wage rate in many cases). Thus, the availability of good quality raw material free from adulterants, in adequate quantity and at suitable prices has been one of the chronic problems of the industry. The demand for these materials is said to be increasing year after year.

However, no efforts appear to have been made either by the government or by the industry to assess the extent of the area under these crops (both in the wild and cultivated form), the present supply position and also the demand from the existing pharmacies. There are around 10,000 licensed pharmacies practising the Indian system of medicine today, whose requirements have to be met.

In addition, there are thousands of local vaidyas, herbal healers, monks, bone-setters and tribal doctors who practice this system of medicine. Many of them depend on retail outlets for their supply of raw materials.

So, we can see that there is a tremendous demand for raw material from all the four segments of the drug industry: *(i)* Plant drugs for Indian systems of medicines covering the Ayurvedic, Unani and Siddha systems, *(ii)* Over the counter non-prescription items involving plants, extracts and Galanicals, *(iii)* Essential oils and *(iv)* Phytophar-maceuticals. Some of the medicinal plants growing in the wild are becoming extinct on account of destructive collection techniques, particularly when the whole plant is uprooted. Sources in the indus-try mention large-scale use of adulterants and other substitutes which reduce the potency and efficiency of the formulations.

Even certain crops which are commonly found like goosebery, neem, clove, nutmeg and dry ginger are also not available in adequate quantities to meet the requirement of the user-industry. The problem is compounded by the fact that the existing land ceiling norms prohibit the user-industry from owning large areas of land and cultivating medicinal crops for their own use.

(b) World Status

A vast majority of the world population today are finding them-selves unable to afford the products of the western pharmaceutical industry, and they have to depend mainly upon the use of traditional medicines. This reality has been recognized, documented and compiled by the WHO in an inventory of medicinal plants numbering over 20 000 species. The world population is expected to touch 750 crores by the turn of the century, and this will further escalate the health budget especially those of the developing countries. The developing countries spend roughly 40–50% of their total health budget on drugs and, as a strategy to reduce the financial burden on developing countries, the WHO encourages, recommends and provides for the inclusion of herbal medicines in natural health care programmes. Such herbal medicines are easily available at a cheaper price for the common man, they are time-tested and considered safer than some of the modern synthetic drugs. Of late, there has been a resurgence of interest in herbal medicines in the Western European countries. This is evident from the fact that 2 volumes of the *British Pharmacopoeia* were published and 33 million US dollars worth of literature on herbal products was sold in the USA in 1990. Japan has recently adopted a health policy based on phytopharmaceuticals to be implemented in the next five years.

The market for herbal medicines in the developed countries is growing at a faster rate than other pharmaceutical products. Some of the reasons are:

i. The realization that allopathic drugs have harmful side-effects.

ii. Allopathic medicines are said to be ineffective against many chronic diseases like cancer. Moreover, many people suffering from diabetes, arthritis, respiratory diseases, skin ailments, gastric problems, jaundice, etc. are said to be turning more and more to Ayurveda and Unani for permanent cures.

iii. Herbal medicines are comparatively less expensive.

iv. The western medical profession has begun to acknowledge the value of herbal medicines. This also explains the fact that many purely allopathic units have also adopted the use of formulations which include natural herbal drugs.

It is estimated that in 1980, the sales of herbal medicines in European community accounted for US $ 2.25 billions (Rs. 7000 erores). This is only 3% of the European pharma market estimated at US $ 66 billion (Rs. 211000 crores). In the EC market for phytomedicines, Germany has the largest share at US $ 1.5 billion (Rs. 4800 crores), i.e. 2.2% of the total. The size of the French market for herbal medicine is $ 0.21 billion (Rs. 670 crores) which amounts to less than 1 % of the EC market. The UK market is $ 425 million (Rs. 1300 crores).

The share of the Indian market is negligible, notwithstanding the fact that we have a varied emporium of medicinal plants and are the exporters of over 200 major drugs and pharmaceutical.

Export of Medicinal Plants from India

India has been a traditional exporter of medicinal plants for the past several decades and ranks as one of the foremost supplier of medicinal plants in the world. The annual export of vegetable drugs, from India has been on the increase and during the year 1995–96 it has gone up to Rs. 189.30 crores, besides the export of Rs. 70.03 crores worth of alkaloids.

Among the items, psyllium husks and seeds was the fore-most foreign exchange earner, leaving behind the opium which once upon a time used to be our star item of export followed by senna leaves and pods and ginseng raots. Export of psyllium husk and seeds druing 1995–6 was to the extent of 154.41 crores. Ginseng roots (other) and senna leaves and pods have earned a foreign exchange of Rs. 16.27 and 13.91 crares, respectively. Among the alkaloids, opium and their derivatives and salts have been the other major items of export earning 23.66 crares. Reserpine, pseudoephidrine, ephidrine, other alkaloids and salts, salts of other derivatives of quinine, nicotine sulphates, neem oil, formulations of other alkaloids, ergot alkaloids, and atropine sulphate constituted the other items of export valuing Rs. 46.3 crares during the same year.

The total import of these commodities during the year 1995-96 was little over Rs. 43.39 crores. Among the major items imported were vegetable saps and extracts, ginseng, pyrethrum flowers, liquorice roots, agar agar, and chirata.

Future Prospects and Constraints

A comparative analysis of the prospects and constraints of the medicinal plant-based drug industry in our country reveals the following.

Prospects

1. The World Health Organisation (WHO) has emphasized the need for better utilization of the indigenous system of medicine, based on the locally available medicinal plants in the developing countries. In the USA and UK, plant-based drugs are being used in recent years on a considerable scale. The former USSR countries, East European countries and China have adopted an integrated system of allopathic, traditional and folk systems of medicine. During the last two decades, there has been a tremendous transformation of medical systems in the world. Owing to the realization of the toxicity associated with the use of antibiotics "and synthetic drugs, Western countries are increasingly aware of the fact that drugs from natural sources are safer. Therefore, an upsurge the use of products based on plants is expected, especially in the field health care products, where rules and regulations governing them are not stringent.
2. Medicinal plants and their derivatives will continue to playa major role in medical therapy in spite of advances in chemical technology and the appearance of cheap, synthesized, complex molecules from simple ones through highly specific reaction mechanisms. The reactions involved are either difficult or expensive to duplicate by classical chemical methods. For example, in Vitamin A, diosgenin and solasodine of plants, where steric forms are possible, chemical synthesis yields a mixture of the isomers which may be difficult to separate. The product obtained by synthesis may, therefore, be toxic or have a different therapeutic effect than what is obtained in nature.
3. Harvesting of medicinal plants is less costly than artifical drug synthesis. Reserpine is a good example of this. The synthesis of the drug costs approximately \$ 1.25/g, whereas, commercial extraction from the plant costs \$ 0.75/g.

4. The vast range of agroclimatic conditions, varying from alpine/ mild temperate to tropical regions with abundant rains and sun-shine, makes it an ideal place for the luxuriant growth of flora and fauna. India is bestowed with incredible natural plant resources of pharmaceutical value. Despite comprising only 2% of the land mass, India is blessed with 25% of the biodiversity. Over 7,000 species of plants found in different ecosystems are said to be used for medicine in our country. The Indian pharmacopoeia records about 100 medicinal plants and their preparations and most of them are available in India. Out of these, quite a few are also recorded in the pharmacopoeias of other countries of the world, and there is a growing demand for them in the international market.
5. There has been a tremendous upsurge in the demand for phyto-pharmaceutical, raw medicinal herbs and vegetable drugs of Indian origin from the Western nations. There is also an increase in domestic demand for raw material used for perfumeries, pharmacies and biopesticidal units. The demand for traditional herbal drugs is also increasing rapidly mainly because of the harmful effects of synthetic chemical drugs; and also because of an expansion of pharmacies manufacturing natural drug formulations.
6. Our country is said to be the proud possessor of an impressive medical heritage which encompasses various system of medicine, *viz.* Ayurveda, Siddha, Unani and also the Tibetan system of medicine. It has an invaluable treasure trove of various scriptures on diverse medical systems.
7. India is a source of cheap labour and skilled manpower which readily absorbs technological changes and also implements the same.
8. Being strategically located, geographically, India could become a potential supplier of phytopharmaceuticals, alkaloids and raw medicinal herbs for the emerging world market. At present, India is not self-sufficient in pharmaceutical products, and drugs worth millions of rupees have to be imported every year by the pharmaceutical industry in order to meet the national

demand for drugs. In order to check this import, it is necessary that utmost attention is paid towards import substitution by producing the raw material and fine chemicals within the country.

9. In addition, these crops have many virtues, like many of them are hardy and yield reasonably well without much care even on marginal lands. They are relatively free from cattle damage and are pilferproof and, hence, can be profitably grown in areas where stray cattle or wild animals or pilferage is a major problem. As it is, medicinal plants are better earners than many field crops. Since they are new crops, there is immense scope for further improvement in their productivity and adaptability, in order to obtain further increase in returns. They are sufficiently hardy and varying in their requirements to have immense potential for incorporation into various existing cropping patterns and can be grown as intercrops, mixed crops and undercrops.

Constraints

Although India is a leading exporter of medicinal plants in the world, the rate of growth of these crops in relation to their economic prospects is not at all satisfactory. The reasons for this apparent backwardness are many and varied.

1. So far, there has been no organized research set-up to continually recharge scientific inputs in order to make their cultivation not only economically viable but also more profitable, so that they can claim their due share in the cropping systems of the country.
2. India is one of the few developing countries which has initiated institutions like the Centre for Medicinal and Aromatic Plants (CIMAP); the Regional Research Laboratories (RRL), at Jammu, Bhubaneshwar and Jorhat; the All India Co-ordinated Project on the Improvement of Medicinal and Aromatic Plants of the ICAR; National Botanical Gardens, Forest Research Institutes, State Cinchona Directorates in Tamil Nadu and West Bengal and, in addition, many Agricultural Universities have

been working on the technological advancement' of this group of crops. However, except for CIMAP and few other centres under ICAR who have taken over a few selected crops for improvement, the replenishment of renewable inputs like quality planting material of improved varieties, developing extension literature, organizing training and quality testing, are very limited.

3. On the marketing side also, it is not much better. The prices of these crops are largely quality based. Lack of testing facilities at the procurement and trading centres together with unscrupulous market handling, have resulted in wide fluctuations in prices, often going down to unproductive and unrealistic levels. Thus, speculative trade has been one of the most serious deterrents to the development of this enterprise.
4. The systematic cultivation of a few medicinal plants has been found to be a discouraging enterprise, mainly because of the uneconomical price they command. For example, the sale price of *Phyllanthus amarus* is as low as Rs. 101kg, making it a commercially unviable proposal. There is a need for the user industry to come forward and ensure that the cultivated product is going to be homogeneous, in comparison to those collected from natural sources, where there is bound to be a lot of variation.
5. Besides, although most of them are, industry-oriented crops, the pattern of land-holdings does not lend itself to commercial cultivation on an extensive scale. In the case of a few plants, *viz.,* amla, asoka, arjun, bael, nutmeg, neem, the cultivation involves a long gestation period due to which many people avoid growing them.
6. Unstable market conditions have also kept farmers away from taking up cultivation of these crops. The prices of certain crops like *Holostemma annulare* fluctuate greatly; the price of the crop/kg ranged from Rs.70/kg in 1993 to Rs. 240/kg in 1990.
7. In the phytopharmaceutical industry, presently, no quality standards have been fixed, either for the raw material or the

final product and, as such, one finds wide variations in the quality specifications.

8. Problems in the identification of medicinal plants have led to the use of adulterants. Physical verification is also a difficult proposition, mainly because the plant part used in many cases like the barks, roots, etc. show close similarity. The only way to check adulterants would be by chemical examination.
9. A number of medicinal plants presently being used have not been adequately studied with respect to their required package of practices or agrotechnology.
10. The supply of raw materials for the phytopharmaceutical industry is virtually monopolized. It is found that supply and price patterns are often determined by the minor forest produce contractors/collectors.
11. In a number of cases, the produce has to be used fresh for which instant transportation is a must, and in many cases it cannot be stored for long periods as this would entail fumigation which, at times, results in chemical contamination of the raw material and eventually the final product, because of its residual effect. Generally, the maximum period for which plant material should be stored is around 5–6 months and no more.

To overcome these constraints, it is necessary to organize the cultivation of medicinal crops on an area-wise basis and organize their marketing on the similar lines as that of other cash crops like coffee, tea, cardamom, to boost their production and returns. In fact, in some states like Tamil Nadu there are some co-operatives which are exclusively engaged in the procurement and marketing of these crops.

1. India, with its vast biodiversity and potential for commercial exploitation, could become a world leader in the supply of raw material for the phytopharmaceutical industry.
2. By drawing up a comprehensive strategy for the cultivation and conservation of medicinal plants in league with the forest departments, many threats outlined earlier could turn into opportunities for successful commercial exploitation without

tampering with the interests of the communities involved in the collection of medicinal plants.

The introduction of medicinal plants into the cropping patterns of farming communities, especially in dryland and watershed areas could provide a strong thrust to the need for soil and water conservation, as well as provide returns and indirectly help in ex *situ* conservation of these crops to a significant extent.

Ashwagandha

Plant Profile

Family	:	*Solanaceae*
English name	:	Winter cherry
Indian name	:	Ashwagandha, Varahakarni (Sanskrit), Askandhatilli, Aasgandh, Punir (Hindhi) Hiremaddinagida, Panneru, Ashwagandha Kiremallinagida (Kannada), Amukkire (Tamil)

Species and varieties	:	*W. somnifera* *W. congulans* Jawahar Asgandh-20
Distribution	:	India, North Africa
Uses	:	Drugs

Ashwagandha, Indian ginseng, Winter cherry, Ajagandha, Kanaje Hindi, Amukkara in Tamil and Samm Al Ferakh, is a plant in the *Solanaceae* or nightshade family. It grows as a short shrub (35–75 cm) with a central stem from which branch extend radially in a star pattern (stellate) and covered with a dense matte of wooly hairs (tomentose). The flowers are small and green, while the ripe fruit is orangered and has milk-coagulating properties. The plant also has long brown tuberous roots that are used for medicinal purposes. It is cultivated in many of the drier regions of India such as Manasa, Neemuch, and Jawad tehsils of the Mandsaur District of Madhya Pradesh, Punjab, Sind. and Rajasthan.

Claimed medicinal uses

Ayurveda

In Ayurveda, the roots of W. somnifera are used to prepare medicinal Ashwagandha. It is claimed to possess aphrodisiac, sedative, rejuvenative and life prolonging properties. It is traditionally used to treat the following symptoms and conditions, although there are few scientific studies of the health benefits of Ashwagandha:

- chronic fatigue
- dehydration
- bone weakness
- muscle weakness and tension
- loose teeth
- thirst
- impotency
- premature ageing
- emaciation
- debility
- constipation
- senility

- rheumatism
- nervous exhaustion
- memory loss
- neurodegenerative disorders
- spermatorrhoea

The berries can be used as a substitute for rennet, to coagulate milk in cheese making. The berries and leaves are traditionally used a topical treatment for tumors and tubercular glands, carbuncles and ulcers. While Ashwagandha is claimed to have a wide variety of health benefits, there have been few clinical trials to test these claims. Studies completed so far support that W. somnifera may possibly be beneficial for:

- easing drug withdrawal symptoms
- reducing anxiety
- reducing arthritis pain in the knee

In addition, there are registered clinical trials in progress to determine if W. sominifera is useful for treating:

- Tuberculosis
- Parkinsons Disease
- Bone Cancer
- Bipolar disorder
- Diabetes
- improve the well-being of the elderly and breast cancer patients undergoing chemotherapy

Naturopathy

In a randomized control trial examining the use of naturopathic care for anxiety, a "naturopathic care" group who received *W. somnifera* (along with "dietary counseling, deep breathing relaxation techniques, a standard multi-vitamin") showed significant improvements in anxiety (as measured by the Beck Anxiety Inventory) compared to a psychotherapy group (who received "psychotherapy, and matched deep breathing relaxation techniques, and placebo") after 8 weeks. No attempt was made to examine the differing contributions of *W. somnifera*, dietary counselling and a "standard multi-vitamin" in the first group, or psychotherapy in the second group, toward the outcomes for each group.

Babchi

Plant Profile

Family	:	Papilionaceae of Leguminaceae
English name	:	Babchi seeds
Indian name	:	Vakuci, Bakuci (Sanskrit), Bakuci Babchi (Hindi), Bavanchigida (Kannada) Karpogam, Karpokkarisi (Tamil)

Species and varieties	:	Psoralea corylifolia Linn.
Distribution	:	India
		Tropical/Subtropical regions of the world
Uses	:	Drugs

Soralea vorylifolia Linn., commonly called Babchi, is a medicinal plant belonging to the subfamily Papilonanceae of the Leguminaceae family. The seed is surrounded by a sticky, oily pericarp which contains coumarins, of which Psoralen and Isopsoralen are therapeutically important. Besides treating psoriasis, psoralin is being investigated as a cure for several diseases including AIDS. It is also used in the treatment of intestinal amoebiasis and the healing of wounds and ulcers. There are several reports in literature on the anti-microbial, antifeedant and insecticidal activities of babchi suggesting other possible uses.

Distribution

The genus Psoralea, is widely distributed in the tropical and sub-tropical regions of the world. Out of the four species occurring in India, the seeds of Psoralea corylifolia Linn., alone are used in medicine. This species has a large distribution in the South-eastern districts of Madhya Pradesh and Uttar Pradesh from where commercial collections are being made. In addition, it has a wider distribution in parts of Rajastan, Andhra Pradesh, Bihar and Gujarat and related semi-tropical grasslands in the country.

Description of the Plant

Babchi is an erect, annual herb, that grows 30–60 cm tall under natural conditions and up to 160 cm under cultivation. The plant branches profusely; its stem and branches are covered with conspicuous glands and white hairs. It bears simple leaves, broadly elliptic, rounded and mucronate at the apex. The axillary, solitary Inflorescence (raceme) comprises 10 to 30 flowers with a hairy pedicle. The calyx is 3–4 mm long and hairy outside; the corolla is bluish-purple, nearly twice as long as the calyx (7–10 mm), zygomorphic, with depending imbricate aestivation; its 'standard' is 6 mm long, orbicular, clawed and glabrous. The stamens, numbering

10, are diadelphous; the carpel is small and unilocular. It bears a single-seeded pod which is indehiscent and the pericarp is usually found adhering to the seed.

Babchi is a self-pollinated crop. The initial flower-bud opens in about 17–18 days. The flowers open in two flushes, once early morning and again in the evening. Anther dehisces about one hour before the stigma becomes receptive. Its chromosome number (2n) is 20 and 22.

Soil

In nature, it is found growing on a variety of soils ranging from sandy medium loam to black-cotton soils. However, sandy loam soil with good organic matter is the best for its growth and yield. It also tolerates a wide variation in soil pH, but too acidic and too alkaline soils should be avoided.

Climate

The crop prefers dry tropical regions with a comparatively warmer type of climate for its successful growth and yield. It is generally grown as a rainy season crop with the onset of the monsoons. It is a hardy plant and grows well in areas with low to medium rainfall during the summer months. Under Bangalore conditions, the crop sown in the month of June has yielded the best.

Land Preparation and Sowing

The field should be prepared to make good seed-beds, by light ploughing followed by two harrowings before the onset of the monsoon season. The seeds are dibbled in rows, preferably 45–60 cm apart, keeping plant-to-plant spacing around 30–45 cm, depending upon the fertility of the land. A seed rate of 7 kg. will be enough to cover a one hectare area.

Cultivation

Propagation

Babchi is propagated through seeds. The seeds have a problem of dormancy due to the hard seed-coat and germination is invariably poor (5–7%). The germination percentage can be improved by breaking the dormancy by the mechanical puncturing of seed coverings or by treating the seeds with sulphuric acid for 50 minutes. A pretreatment of seeds in 1% sulphuric acid has been found efficacious; the sulphuric acid should be washed off by repeated washings in water before the seed is sown.

Manure and Fertilizers

The crop responds well to the use of organic and inorganic fertilizers. It has been noted that the basal application of Farmyard Manure (FYM) at 20 t/ha gives a good initial growth and increases the seed yield significantly. Besides, a fertilizer dose of 100 kg N, 60 kg P and 50 kg K/ha is recommended. Of this, half a dose of N and a full dose each of P and K is given as a basal dose, while the remaining 50 kg N/ha is applied in one dose as a top-dressing after 45 days of sowing.

Irrigation

Since it is mostly grown as a rainy season crop, the irrigation requirement of babchi is moderate. However, after the rainy season is over (June-September), the crop may be irrigated fortnightly. About 6 to 8 irrigations may be given until the crop is finally harvested.

Intercultural

The field must be kept free from weeds by regular weeding as and when required. In all, about 2 to 3 hand-weedings are necessary during the early period of growth after which, the plants cover the ground densely providing practically no opportunity for the growth of weeds.

Pests and Diseases

The plant is sufficiently hardy and free from pests. However, under Bangalore conditions, the Lepidopteron insects, viz. the leaf folder and two defoliators have been found to cause considerable damage to this crop. These insects can be effectively controlled by spraying monocrotophos @ 1.5 ml/l of water.

Among the diseases, the powdery mildew is the only disease of a serious nature noticed on this crop. This disease can be controlled by spraying wettable sulphur @ 3 g/litre of water at weekly intervals.

Harvesting and Yield

The crop takes about 7–8 months from the time of sowing to reach maturity. As the seeds continue to mature over the growing plants continuously, the seed-picking will commence from mid-December onwards to the end of February. In all, 4 to 5 pickings are usually taken. At maturity, the single-seeded fruit turns brownish-black and emits a mild odour.

An average yield of about 2 t/ha of dry seeds may be obtained.

Procedure for the Estimation of Psoralen from the Seeds

It has been reported that leucoderma of non-syphilitic origin can possibly be treated with seeds and seed preparations. The therapeutically active principle is reported to be psoralen. Encouraging results have been reported with the oral administration of psoralen on patients below twenty years of age with leucoderma of recent origin.

The seeds contain a brown fixed oil (10%), a dark brown resin (8.6%), psoralen/isopsoralen (1%), an essential oil (0.05%), a non-volatile terpenoid oil, a monoterpenoid phenol and raffinose. A large number of procedures for the extraction, isolation and determination of psoralen/isopsoralen have been reported. Some of them have been covered by Indian, British and US patents. The basic principle involved in all the processes is to liberate the furocoumarins, present in the seeds as glycosides of the corresponding coumarinic acids, by hydrolysis. The following procedure was standardized for the estimation of psoralen/isopsoralen from the seeds of Psoralea corylifolia.

To estimate the psoralen content, the seeds of Psoralea corylifolia are defatted with petroleum ether, dried and coarsely powdered. The seed-powder (2–5 g) is soaked in 95% ethanol, kept overnight and then percolated through 95% ethanol. The extract is then concentrated under reduced pressure to yield a thick, syrupy residue, which is then extracted with solvent ether to separate the non-glycosidic fraction (1). The residue, which is insoluble in solvent ether and which contains psoralidin, is hydrolyzed by refluxing with alcoholic HCI. The alcohol is removed under reduced pressure and the hydrolysis mixture extracted with solvent ether, washed with water, dried over sodium sulphate, filtered and the ether is distilled off (2). The residues (1) and (2) are passed over a silica gel column and eluted with 1 % methanolic chloroform. Different fractions are monitored on TLC and all the fractions containing psoralen are combined, the solvent is distilled off under reduced pressure and the residue is washed with petroleum ether (60–80°C). It is then cooled in the refrigerator for two days and filtered and washed with a little cold petroleum ether. Psoralen, thus obtained, is dried, weighed and the percentage is calculated on the basis of the weight of the seeds.

Belladonna

Plant Profile

Family	:	Solanaceae
English name	:	Belladonna, Deadly Nightshade
Indian name	:	Angurshefa, Luckmuna, Kotu, Phaphra (Hindi)
Species and varieties	:	A. belladonna Linn. (Deadly Nightshade)
Distribution	:	A. acuminata Royle (Indian
	:	Atropa)
Uses	:	Drugs

Belladonna (Atropa), belonging to the family Solanaceae, comprises four species of medicinal plants. Out of these four species, the commercial drug is obtained from the leaves, flowering tops and roots of A. belladonna Linn. commonly called 'Deadly Nightshade' and A. acuminata Royle, often referred as 'Indian Atropa' or 'Indian Belladonna'. The genus name Atropa has been derived from 'Atropos' the Greek goddess of death, whereas, the word 'belladonna' is derived from the Latin word 'Bella' meaning beautiful and 'Donna' meaning woman. Although in most of the countries of the world the drug is obtained from A. belladonna, in India all the Belladonna comes from plantations which are a mixture of A. belladonna and A. acuminata.

The leaves and roots of belladonna contain tropane alkaloids whose concentration varies from 0.13 to 0.7% (the average is 0.45%). A. belladonna contains alkaloids, chiefly l-hyoscyamine and small amounts of atropine, its optically inactive isomer, and small amounts of hyoscine (scopolamine). In addition, there are small quantities of volatile bases, such as pyridine and N-methyl-pyrroline. The leaves also contain β-methyl esculetin (scopoletin), succinic acid, asparagin, etc.

A. acuminata leaves contain 0.45% hyoscyamine and non-alkaloidal volatile bases not greater than in A. belladonna. The roots of A. acuminata contain 0.20-0.8% hyoscyamine (0.47% is the average) and non-volatile alkaloids (0.4%).

Belladonna leaves were included in the British pharmacopoeia in 1809. The roots, however, were accepted as a drug only in 1880. Belladonna leaves are widely used for the manufacture of tinctures, extracts and plasters. The drug serves as an anodyne, sedative-stimulant, antigalactogogue, antidiuretic, mydriatic or anti-asthmatic, antispasmodic and anti-inflamatory. It is also used for relief from whooping cough, in the treatment of renal and bilary colic, in stomach disorders, to stop sweating, in opthalmology to cure gastric ulcers and as an antidote for depressing poisons such as opium, muscarine and chloral hydrate. The roots are primarily used in the external treatment of gout, rheumatism and other afflictions. It is also used in the treatment of epilepsy, Parkinson's disease and bradycardia.

Origin and Distribution

A. belladonna is indigenous to Southern and Central Europe and naturalized in Southern England. It is cultivated for its drug in Central Europe, England, the USSR, the United States and Northern India. A. acuminata is found in a natural state in the Western Himalayas extending from Kashmir to Simla and the adjoining areas of Himachal Pradesh. Its chief habitat in Jammu and Kashmir state are the forests of Sindh, Jhelum, Udder and Chenab Valleys. It is also cultivated in some places in the Himalayas.

Description of the Plant

A. belladonna (2n = 72, a hexaploid) is an erect, herbaceous plant, 150 cm high, with a long, cylindrical, branched, fleshy, creeping, perennial root, varying from 2.5 cm to several centimeters in width, yellow-brown externally, white within. The plant is shrub-like with spreading slender, cylindrical, sometimes slightly downy, often purplish, branches. The leaves are alternate, but on the upper branches, each leaf is usually accompanied by a small, stipule-like leaf. The blade is ovate, pointed, entire, soft, with conspicuous veins and frequently with a purplish mid-rib which may be hairy beneath; 7.5–22.5 cm long, 5.0-10.0 cm wide and petioles of 0.6–3.75 cm. The flowers, borne singly in the leafaxils, are large, bell-shaped, the upper part is brownish-violet or purple or golden-brown with violet-brown veins. The fruits are globular, slightly bilobed, shiny purple-black, subtended by an enlarged calyx-lobe. The fruit is a berry and it reaches 1.25 cm in diameter and is two-celled and contains many kidney-shaped seeds embedded in a pulp which is somewhat sweet and yields a purple juice. In the variety Lutea, the leaves and stems are yellow-green.

A. acuminata is a tall perennial herb, 70 to 150 cm in height with a dichotomously branched stem. It closely resembles A. belladonna. It has a large taproot with lateral rootlets 6 inches or more in length and of 1–2 cm thickness. The leaves are green or olive-green in colour, 7.5–21.5 cm long and 4–9 cm in breadth. The flowers are brownish-yellow with greenish veins and are borne singly or in groups of two to four. The fruits are globular and purple-black in colour.

Varieties

There are no named varieties reported under this crop but the workers at the Regional Research Laboratory, Srinagar, have selected plants containing 0.60% alkaloids in A. belladonna.

Soil

Belladonna grows well in slightly acidic, deep, fertile soils of medium texture, which are rich in humus. Heavy clay soils which are water-logged should be avoided.

Climate

Belladonna is a temperate plant. It behaves as a perennial in temperate climates and gives maximum herbage and alkaloid yield. In subtropical areas it can be grown as a winter crop. However, the plant behaves as an annual as it dies during the summer months and hence the yield is poor.

Land Preparation

The land is prepared well and brought to a fine tilth with the help of tractors and organic manure at the rate of 25 t/ha, to begin with, is mixed with the soil.

Cultivation

Propagation

The crop is propagated through seeds extracted from the berries, usually collected from September to November. About 4 kg of seeds will be required to raise seedlings for one hectare of land. The germination is very poor and it takes 3–6 weeks for the seeds to germinate. Hence, the seeds may be treated with 80% sulphuric acid at the time of sowing for 2 minutes to improve germination. Stratification of seeds on wet sand under 5 to 12°C or–2 to 2°C for 10–40 days has given a high percentage of germination in other countries.

Vegetative propagation by root-division from 3–4 year-old belladonna plants is also suggested by the Drug Research Laboratory, Jammu and Kashmir.

Nursery Raising

Belladonna has been cultivated by direct sowing, but raising them in the nursery gives the best results. The nursery may be raised from the second week of May to the end of autumn (September to October) and in summer under sufficient shade. Before making the nursery beds, the land should be ploughed well so as to give a fine tilth. Raised beds of 3m × 1m size may be made, surrounded by drainage and irrigation channels. Well decomposed FYM or sheep manure should be mixed in the top 10 cm of the soil. Seeds pre-treated with fungicides like dithane M–45 or Agallol (10 g/kg of seeds) may be mixed with fine soil in a 1:40 ratio to ensure uniform sowing and broadcast (@ 200 g seeds/m^2) in the nursery-beds and covered with a layer of FYM and then with straw. The beds must be watered immediotely after sowing with a watering can and daily thereafter. The seeds normally germinate within 3 weeks with a good germination percentage, and the seedlings will be ready for planting in the main field when they attain a height of 15–20 cm after about 8–12 weeks.

Broadcasting

About 20 kg seeds are required for sowing one hectare of land by broadcasting.

Transplanting

The ideal time for planting in the field is early spring (March-April) or autumn (October-November). Generally, seedlings raised in autumn are planted in spring and those raised in summer are planted in autumn. Seedlings are planted at a spacing of 50–60 cm in rows about 60–70 cm apart. It is always safer to plant the seedlings on raised beds with 1 m-wide strips or ridges as it avoids water-logging and facilitates irrigation. The field may be irrigated immediately after transplanting.

Fertilizer Application

Belladonna is an exhausting crop, hence a basal dose of 25–40 kg N, 40–60 kg P_2O_5 and 30-50 kg KO/ha is supplied. An additional dose of 60–80 kg N is applied in 3–4 split doses as a top-dressing at monthly intervals after every harvest. The application of heavy doses of nitrogen gives a good crop of leaves and also increases their alkaloid content.

From one year after planting, a basal dose of nitrogen and phosphorus should be repeated every year in the beginning of the spring season (March-April), which is followed by a top-dressing of nitrogen. Micronutrients have been found to affect the growth. It is better to irrigate the field after each application of fertilizers, if there is no rain.

Irrigation

Belladonna has a high water requirement and it should be irrigated frequently, once in 10–15 days during the dry period. Normally, 6–7 irrigations are required during the dry months. Care should be taken to avoid water-logging.

Interculture and Weeding

The crop should be kept free from weeds by frequent weeding and hoeing. One hand-weeding is necessary when the plants are young. A tractor-drawn cultivator or hoe may be employed for later weeding.

Pest and Diseases

Pests

Among the insects, cut-worms (Agrostis flammetra) cause consider-able damage to young plants. The other insects noticed on this crop are the potato beetle and flea beetle. The attack is severe in the early summer months (April to May), when the seedlings and young plants are cut at the ground-level. The pests can be controlled by treating the seed bed with Aldrin just before planting, and treating the soil with a solution of Chlordane (1:19) 2 weeks after germination of the seed.

Diseases

A species of Phytophthora, a root fungus, causes havoc in European belladonna plantations. Whereas, in India, Pythium butleri causing root-rot is the only serious disease-causing organism to the bella-donna plantations. Plants at every stage of growth are affected. In young seedlings, the fungus causes pre- and post-emergence damping-off of the seedlings.

This can be controlled by fumigation of the soil with methyl bromide and treating the seeds with Agallol or Captan. Older plants, when affected, will wilt. After a 3-year crop of bella-donna, a crop rotation may be followed by planting the field with a crop like rye or wheat which should be followed by a 5-year crop of pyrethrum. The fungus Ascochyta atropae causes leaf-spot disease, whose symptoms are the appearance of small, roundish, brown spots on the leaves. In severe incidence, blighting of leaves is noticed. The disease can be controlled by spraying Blitox. Pythium spp. is reported to cause damping-off of the seedlings, for which drenching of the field with 0.1 % Dithane M–45 is re-commended. Belladonna mottle virus has been observed on A. belladonna, for which the vectors are not known. Uprooting the plants and burning them in isolation or burying them deep are the control measures suggested.

Harvesting

The first harvest of leaves is available three months after planting. The initial yield is poor because of the fewer number of branches; the yield of leaves increases after the first cutting. Harvesting should be done as soon as the shoots start flowering, as it is the period when the alkaloid content is higher. The leaves are cut with the help of a cutter (pruning scissors) about 30 cm above ground-level as against 7.5 cm above ground-level, done earlier. During the first year, 3 to 4 crops of leaves are obtained. Usually, the crop will be retained for 3 to 4 years, after which new planting may be taken up.

Drying

The leaves are dried immediately after harvest under the shade or sun or on wire-racks, taking care to see that the dried leaves remain green. The leaves may also be dried with artificial heat, with or without fans for air circulation. The harvested leaves are spread in thin layers and turned frequently while drying. The woody stems are discarded before drying. Caution: Prolonged drying reduces the alkaloid content.

The roots are harvested after 3 years. After harvest, they are washed, cut into 4-inch long splits, length-wise if thick and shade or sun-dried. The dried crop is pressed hydraulically into large hessian cloth bags and stored in a cool, dry place away from light, or are transported right away to processing units.

Yield of Herb and Roots

During the first year, an average of 6 q/ha of dry herb is obtained. The yield increases to 12 to 15 q/ha during the 2nd and 3rd year. The yield of dry roots will vary from 1.7 to 3.35 q/ha.

Buckwheat

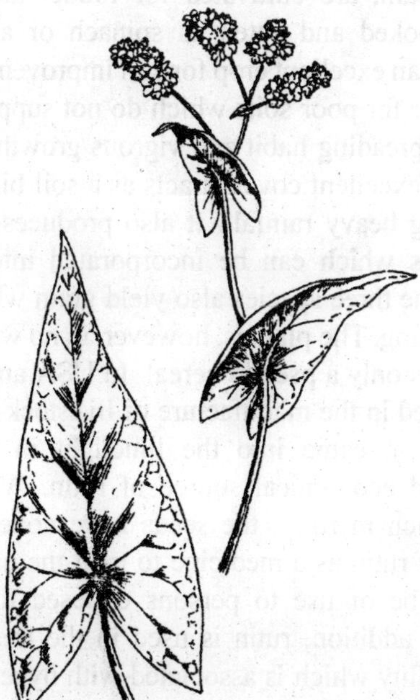

PLANT PROFILE

Family	:	Polygoneaceae
English name	:	Buckwheat
Indian name	:	Kutu (Hindi)
Species and varieties	:	Fagopyrum esculentum Moench, F. cymosum F. tararicum
Distribution	:	India, Central Asia, Europe, France Russia, USA

Uses : Drugs, Food and vegetable crop
Livestock and poultry feed

The genus Fagopyrum, belonging to the family Polygoneaceae, has three important species, of which Buckwheat (Fagopyrum esculentum Moench) is an important medicinal plant, commercially grown for Rutin. The other two species of this genus, F. cymosum and F. tararicum, are cultivated for fodder and some-times their leaves are cooked and eaten as spinach or as amaranthus. It is considered as an excellent crop for soil improvement, particularly for virgin lands or for poor soils which do not support any other cereal crop. By its spreading habit and vigrous growth, it smothers weeds and forms an excellent cover. It acts as a soil-binder preventing soil erosion during heavy rainfall. It also produces a large quantity of green biomass which can be incorporated into the soil as green manure. All the three species also yield grain which is used as flour for bread-making. The plant is, however, in no way related to the real wheat and it is only a pseudo-cereal. In USA and Russia, the grains are mostly used in the manufacture of livestock and poultry feed.

Buckwheat came into the limelight in 1946, as the most promising and economical source of rutin. While the leaves and flowers are rich in rutin, the seeds have none. With the growing importance of rutin as a medicine to treat the harmful effects of X-rays, it may be of use to persons exposed to dangerous atomic radiations. In addition, rutin is used in the treatment of increased capillary fragility which is associated with hypertension a condition which sometimes results in the bursting of blood vessels in the brain, leading to apoplexy or retinal haemorrhage and causing partial or complete blindness. It is useful in a variety of haemorrhagic conditions which include certain types of purpura, *i.e.,* bleeding from the kidney, heriditary haemorrhagic telangiectasia and haemophilia. It has been used with success as a prophylactic treatment against gangrene due to frostbite. Rutin injection with certain miotics is useful in glaucoma, and in combination with dicummarol in the treatment of retinitis. It also prevents the weakening of capillaries due to drugs such as salicyclates, arsenicals, thiocyanates, sulphadiazines and gold salts. The usual dose is 20–30 mg repeated 3–4 times/day for several weeks.

Origin and Distribution

Buckwheat is said to be a native of Central Asia and is cultivated in various countries as a food or fodder crop. In Europe, Russia, USA and France it constitutes one of the main food crops. In India, it is grown as a grain crop in the higher altitudes (600–3 650 m) of the Himalayas (from Kashmir to Sikkim) and in the Nilgiris, but more often as vegetable crop in the lower mountain ranges.

Description of the Plant

It is an annual herb growing up to 80–160 cm in height. The leaves are alternate, acute and 2.5–7.5 cm long; flowers are produced in auxiliary or terminal cymes and are pinkish-white in colour, fragrant, diamorphic and highly self-sterile. The fruit is of 3 colours, ranging from silvery-grey to brown or black.

Soil

The crop can be grown on a variety of soils, but it thrives best on well-drained sandy or loamy soils. It does not grow well on poorly-drained, heavy or acidic soils.

Climate

A moist, cool climate is ideal for its optimum growth. It is a quick growing crop, requiring 7 to 9 weeks from seed to seed. A monthly average temperature of 18 to 30°C seems to be satisfactory. As it is a photosensitive crop, flowering is very erratic and sometimes plants as small as 6 to 8 cm are seen flowering, hence the period of cultivation should carefully be chosen for its optimum production. In most places in India, early September to early December seems to be the best, during which period at least two crops can be cultivated in the same field. A crop sown in the last week of October takes 46 days to reach maturity with an appreciable plant height of 47 cm. Crops sown four weeks earlier and later than the above period of sowing, resulted in poor total fresh and dry weight of herb and crude rutin content. Therefore, it is suggested that when buckwheat is intended to be introduced into a new region or area, field trials should be conducted with staggered planting at narrow intervals to find the most suitable season for successful crop production.

Cultivation

The crop being a short duration, medium-sized annual and sensitive to water-logged situations and a poor competitor with weeds, thorough prepration of the land so that it is free of weeds is necessary. The land must be thoroughly ploughed 3 to 4 times and brought to a fine tilth by harrowing. The field is divided into plots of convenient size (4 to 6 sq m) for the proper distribution of irrigation water. The soil within the plots is levelled. F. esculentum requires a seed rate of 60–80 kg/ha whereas, for the other two varieties, 40–50 kg seeds/ha is sufficient. The seeds are line-sown with a spacing of 20 cm between rows and 10 cm within each row. The most favourable temperature for germination is found to be 35°C. In the crop sown between December and January, there are chances of delay in germination due to frost.

Manures and Fertilizers

The field is moderately manured with well-rotted FYM. An application of 50 kg N and 40 kg each of P and K/ha is found to be adequate for a good crop. Half the quantity of nitrogen along with the full dose of phosphorus and potash are applied as a basal dose. The remaining half-dose of nitrogen is applied 4 to 5 weeks later.

Irrigation

It is generally grown as a rain fed crop with one or two protective irrigations, if required.

Interculture

The crop is kept weed-free by frequent hoeing and weeding in the early stages of growth.

Pests and Diseases

Insect Pests

No major insects affect this crop

Diseases

A few bacterial and viral diseases are noticed on this crop and are not of a serious nature. However, leaf blight which appears during flowering or even at earlier stages of growth, causes some damage to the crop. Initial symptoms start with minute spots on leaves which quickly develop and coalesce to from bigger patches, which turn into blight and finally result in leaf-fall and the drying up of the plant. The leaf blight disease can be controlled effectively by spraying 0.2% copper oxycholoride, whenever the disease is noticed.

Harvesting and Yield

The best stage for harvesting is 40 to 50 days after sowing and when the plants begin to bloom. Harvesting should, however, be done well before grain-formation, if the crop is intended for rutin production.

Delayed harvesting causes a sharp decline in rutin content from 6.0 to 3.8%. The period between blooming and seed-setting being short, close observation is required to enable harvesting the crop at the proper stage. The crop at this stage which would have generally attained a height of 60 to 70 cm is cut and harvested above ground-level.

Post-harvest Drying of the Crop

As the stems are hollow and the leaves contain 88 to 90% moisture, sun-drying is slow and may also lead to fermentation and enzymatic action leading to a loss of rutin. Dehydration technique by artificial rapid drying of the leaves is the best suited method. A process for rapid dehydration of leaves using an alfalfa drier has been developed for obtaining leaf-meal rich in rutin and stable under ordinary storage conditions for at least a year.

Yield of Fresh Herb and Rutin

An average yield of 5 000–6 000 kg herb/ha is considered satisfactory. Chemically, rutin ($C_{27}H_{30}O_{16}$) is a flavanol-3-0-di-glycoside, a greenish-yellow, crystalline powder obtained from the leaves and flowers of buckwheat plants. The dry herb yield per hectare is around 500 kg. The leaf to stem ratio is 0.6:i, the crude rutin being 4.9%. Thus, 100 kg of fresh herb, on an average, yields 10 kg of dry herb.

Cinchona

Plant Profile

Family	:	Rubiaceae
English name	:	Quinine bark, Peruvian bark
Indian name	:	Sinkona, Kunayanah (Sanskrit) Kunain (Hindi), Cinkona (Tamil)
Species and varieties	:	C. succirubra C. officinalis

		C. ledgeriana C. robusta C. hybrida
Distribution	:	India, Andes, Bohemia, Bolivia, Burma Colombia, Costa Rica, Eucador, Guatemala Indonesia, Java, Kenya, Panama, Peru Philippines, South America, Sri Lanka, Tanganyika, Tanzania, Uganda, Zaire
Uses	:	Quinine, Anti-malarial drugs

Cinchona Linn. belongs to the genus Rubiaceae which has about 65 species. Among these species, C. succirubra, C. officinalis, C. ledgeriana, C. robusta and C. hybrida am grown commercially for cinchona bark which is the source of quinin a and other anti-malarial drugs. In addition, more than twenty other alkaloids have been isolated from cinchona, of which cinchonidine, quinidine and cinchonine are the most important. The other alkaloids present are quinamine, quinicine (quinotoxinl, cinchonicine (cinchotoxinl, hydroquinine, hydroquinidine, hydrocinchonine (Cinchotine), hydrocinchonidine (cinchonidine), cinchonamine, quinamine, conquinamine, poricine, etc. The alkaloids exist chiefly as salts of quinic and cinchotannic acids, and their relative concentrations vary in different species. The leaves contain 1% total alkaloids. In addition to the alkaloids, the bark also contains the bitter glycoside, a-quinovin, cinchofulvic, cinchotannic and quinic acids, a bitterish essential oil possessing the odour of the bark, and a. red colouring matter. Quinovin, on hydrolysis, yields quinonic acid and mannitan. In individual plants, the total alkaloid content increases up to the age of 8 to 12 years and then begins to decrease. The alkaloids are formed during the descent of the sap, and their concentration is low in twigs and increases down the stem to a maximum in the root-bark. In C. ledgerina, almost 90% of the total alkaloids of the stem-bark is quinine, while of the total alkaloids of the root-bark, only 60% is quinine.

Quinine, as a cure for malaria, has been known of for a long time and, even today, expert medical opinion regards quinine as a safe and sure remedy for malaria. Besides, quinine was in use as an anaesthetic as a substitute for cocaine. Its anaesthetic action is prolonged. It has been used as a sclerosing agent in the treatment of internal haemorrhoids and varicose veins. Quinine protects the skin against sunburn. It is a bitter tonic, stomachic and appetizer. In small doses, in the form of a solution, it is a mild irritant and a stimulant of the gastric mucosa and other mucous membranes. It is a weak uterine stimulant. Quinine and other alkaloids have a number of non-medical uses also. Salts of quinine are employed in beverages, as an addition to hair-oils, as a vulcanization accelerator in the rubber industry, for making polarized lenses and has various other uses in photography and optics.

Origin and Distribution

Cinchona is indigenous to South America, and occurs in the Andes, Colombia, Eucador, Costa Rica, Peru and Bolivia between 10°N and 19°5 at attitudes ranging from 750 to 2 725 m (25,000–9 000 ft).

Some species are also cultivated in Java, Sri Lanka, India, Burma, Tanganyika, Bohemia, Panama, Indonesia, Guatemala, Uganda, Philippines, Tanzania, Kenya, Zaire and Costa Rica. In India, it is confined to some parts of West Bengal, Tamil Nadu (the Annamalais, Naduvattam, Shevaroy Hills, Palani and parts of Tirunelveli) and Karnataka (South Kanara and Coorg) in an area of about 6000 to 8000 ha.

Botany and Description of the Species

Cinchona spp. (x = 17) are evergreen shrubs or trees. The leaves are opposite, simple, entire; the stipules are interpetiolar and deciduous. The inflorescence is a terminal panicle. The flowers are small and fragrant; the calyx is small, united and has pointed lobes; the corolla is tubular with 5 spreading lobes and a frill of hairs along the margins; it is heterostylous. In microstyled plants, 5 exserted anthers alternate with corolla lobes, while the bifid stigma reaches half the length of the corolla tube. In macrostyled plants, the stamens are half the length of the corolla tube and the stigma is exerted. The fruit is a capsule, dehiscing from the base upwards with 40-50 small, flat, winged seeds.

The following are the commercial species.

a. C. succirubra Pav. ex kl. (2n = 34)

It attains a height of 40–50 ft or more. This species has a red bark. The quinine content (0.5 to 1.5%) in the bark of this species is very poor. The species is vigorous and grows well between 660 to 1 800 m (2 000 and 6 000 ft) altitude. The species is possibly a variety of C. pubescens.

b. C. officinalis L. (2n = 34)

This species is well suited for the higher altitudes of South India and flourishes well at elevations of 1 800 to 2 420 m (6 000–8 000 ft). It is a slender tree, 20–30 ft high. It is the source of 'Crown' or 'Loxa bark'. The quinine content of the bark ranges from 1.5 to 2.0%. The species is indigenous to the mountains of Colombia and Peru.

c. C. ledgeriana Moens and Tremen (2n = 34)

This is a weak but fast growing species, attaining a maximum height of about 20 ft, and it can be grown between 900 to 1 800m (3 000–6000 ft) elevations. It is a source of 'Ledger bark' or 'Yellow bark' which is the richest in quinine (4 to 5%); the quinine content is occasionally as high as 14%. This species- is considered either a hybrid between high-yielding varieties like C. calisaya and C. succirubra or a complex hybrid between C. calisaya, C. succirubra and C. lancifolia.

d. C. robusta

The origin of this species is uncertain, but it is presumed to be a hybrid between C. succirubra and C. officinalis. In India, it is grown in the Darjeeling Hills, West Bengal, and can be grown in places with an elevation of up to 1510 m (5000 ft). The average quinine content of the bark is about 2%.

e. C. hybrida

This vigorous growing hybrid is a cross between C. succirubra and C. ledgeriana. The yield of bark is high and the quinine content varies from 2 to 2.5%.

Varieties

Cinchona is a naturally cross-pollinated plant, due to which a great deal of admixture has taken place among the various species and the present stands of cinchona are only crosses between these various species. An outstanding clone-No.701 containing over 12% of quinine sulphate in the bark has been spotted by the Tamil Nadu Forest Department and is propagated on a extensive scale in the Nilgiris and Anamalais.

Soil

Cinchona prefers a light, well drained, virgin forest soil, rich in organic matter with no possibility of subsoil water-logging and with a high moisture holding capacity. It prefers acidic soils (pH 4.6 to 6.5). The calcium requirement of this crop is high.

Climate

C. ledgeriana grows best in tropical climates at altitudes of 1 800 m (6 000 ft). However, the other species grow well in areas with an average minimum temperature of 13.5°C and a maximum of 21°C with a mean relative humidity of 83%.

The growth is very poor at temperatures below 7°C or above 26°C. It grows well in places where the annual rainfall is a little less than 200 mm and is well distributed over at least eight months of the year. Cinchona is susceptible to frost and, hence, is not grown on very high hill ranges.

Land Preparation

For raising cinchona plantations, virgin forest soils are best suited. The forest is cleared and the ground is levelled and dug to a depth of 1½ ft to improve the soil structure and to remove stones, if any. About a fortnight prior to planting, pits of 60 × 60 × 60 cm are dug and filled up with top soil and well decomposed organic matter.

Cultivation

Propagation

Cinchona is propagated both by seeds as well as vegetatively by cutting, stooling, layering, cleft-grafting, vineer or side-grafting and patch-budding. As most of the species of cinchona are highly heterozygous, vegetative propagation is preferred. But, in India, cinchona is propagated by seeds as it is comparatively less expensive.

Propagation by Seeds

Cinchona seeds are small and light and lose viability on storage. The seeds are generally sown during April in sloping beds, 12 × 4 ft, and covered with a thatched roof. The nursery area is prepared in such a way that the top layer, up to a depth of 2–3 inches, is composed of a mixture of leaf-mould and sand in equal proportions, and is carefully pressed by hand so that it is uniformly firm all over. Cattle manure is usually not applied to the nursery beds.

While sowing, the fresh seeds are scattered thickly on the surface and covered with a thin layer of fine sand. The beds are then lightiy watered using a spray can. The seeds germinate in about 20-40 days after sowing. About 50% germination is noticed, out of which only about 10% of the seedlings will be suitable for transplanting after eliminating all the weak, lean and lanky seedlings. The seedlings grown in the nursery can be transplanted in baskets or polythene bags when they are about four months old, with two pairs of leaves. The seedlings will be ready for planting in the main field during mid-May of the succeeding year when they are about 14–18 months old and 30–60 cm in height.

Vegetative Propagation

Among the vegetative methods of propagation, patch-budding, soft terminal cuttings and layering have recorded the best results. A high percentage (85%) of success is obtained in patch-budding in the period from March to the middle of June. Patch-budding is usually done in the nurseries or plantations, in situ on plants which are one to two years old. Even the budding on coppices has given a better growth of budded plants. The cuttings are made to root by cincturing and etiolation. During May-June, the shoots are treated, from which cuttings are taken after 50–65 days and planted in the nursery. In layering, the East Mailing method with some modifications has given good results, each shoot giving annually about 100–200 shoots in 2 or 3 coppicings. C. succirubra is used as stock wherever grafting is done.

Planting

Planting is done before the onset of heavy rains. The soil should be sufficiently moist at the time of planting. The planting is done in open pits at a spacing of 120 × 120cm or 150 × 150cm or dense planting of about 8000 plants per hectare is done. The plants are gradually harvested from 3-5 years, until about 800 plants remain in one hectare after 25 years. Young cinchona plants need shade which is provided by planting shade plants like Alnus nepalensis, Erythrina indica, Albizzia stipulata and Grivellea robusta, 20 ft apart.

Application of Fertilizers

Application of a fertilizer mixture containing N, P and K gives marked results. Liming (dolomite or limestone) is done for the soil if the pH is 5 or lower. Nutrients are supplied @ 115 kg N, 105 kg pp and 115 kg K_2O/ha in the form of triple superphosphate, muriate of potash and ammonium sulphate. The quinine content in cinchona is known to increase with the age of the trees, under favourable nutritional conditions.

Weeding

Weeds have to be removed at regular intervals, particularly in young plantations.

Cover Cropping

Besides the native cover, erotalaria usaramoenis or Leucaena glauca are grown as cover crops in cinchona plantations. When the plantation becomes older the ground is shaded more heavily, hence the original cover crop must be replaced with shade-tolerant species.

Cultivation Operations

The plants should be staked properly in the first three years to prevent their swaying or leaning.

Pests and Diseases

Pests

The grubs of cockchafer beetles cause serious damage to the seedlings in South Indian plantations. The adults of these beetles come to the surface during summer and should be flushed by irrigating the seed-beds with water mixed with crude oil emulsion. The beetles must be then hand-picked and killed.

Crickets also cut the young seedlings and drag them to their haunts in the seed-beds. The haunts must be searched and the insects killed. The application of tobacco decoction with kerosene or phenyl wards off these insects.

The tea-fly, Helopeltis antonii and H. cinchonii, infest the leaves in nurseries and also in the main field. These insects cause leaf-curl by sucking the juice from the tender leaves. If heavy damage is noticed, sodium arsenate mixed @ 28 g with 113 g of molassses in 9 litres of water may be sprayed on the plants for controlling these insects. The leaf bug, Disphinctus humeralis, is occasionally found on tender foliage.

Diseases

Stem blight caused by Sporotrichium and Verticillium spp., seedling blight due to Phytophthora parasitica, root-rot due to Phytophthora cinnamomi, and Sclerotium rolfsii and damping-off due to Pythium vexans and Rhizoctonia solani are some transplant bed-diseases affecting cinchona seedlings. The spread of these diseases can be checked by scooping out the soil in the diseased patches all around, to a depth of 10 cm and throwing it away. The cavity, so formed, may be filled with sterilized, dry soii. The nursery-beds should also be drenched with mercurial fungicide to check the spread of the diseases.

Trunk and branch cankers are plantation diseases which are most prevalent during the rainy season. C. ledgerina and its hybrids are more susceptible to this disease. Dark, shrivelled, brittle patches appear on the stems or branches. The pruning of the diseased parts is recommended. Besides, a physiological canker disease has been reported to affect C. succirubra and C. ledgeriana in Burma. As a result of this canker, the collar of the affected plant swells to about 3 times the normal size and the bark turns dark brown to black. The leaves drop off and those that remain on the plant are much reduced in size. The plants which are badly affected will die, collar canker is said to be caused by too deep planting combined with wind action.

A bark disease and collar rot caused by Fomes noxius and Sclerotium rolfsii, respectively, have also been reported. Die-back or pink disease is caused by Pellicularia salmonicolor. This organism attacks the tender tips of the stems and branches and gradually spreads. The branches should be pruned to prevent the spread of the disease. Besides, the fungus Armillaria mellea causes root rot and Rosellinia spp. causes leaf spot diseases.

Harvesting

By judging the amount of vegetative growth, the trees are coppiced when they are 6 to 8 years old. Coppicing involves prunning the trees at a height of 5 cm (2 inches) from the ground-level. The leftover stump regenerates to produce a large number of shoots, but only two or three of these are retained and allowed to grow further. The rest of the coppices are removed. A second coppicing is done 8–10 years after the first coppicing, where only about 2 to 3 shoots are left to grow further. The plants are uprooted in the 30th year when their vigour declines. The major harvests are obtained at the time of the first two coppicings and only little yield of bark is obtained from the dead and drying trees and prunings. The first set of yields are obtained in the third year after planting. The bark is separated from the coppices by beating it with a mallet and is then peeled by hand or a knife. The peeled bark should be dried immediately to prevent the loss of alkaloids, preferably in the shade. In rainy weather, drying is done in special sheds or by means of artificial heat. In well-established plantations, drying is done in well-regulated ovens. For this purpose, hot air ovens, regulated by 70°C, are employed. The long, strips of bark are cut into small pieces and fed into the upper end of a long, slightly inclined, rotating, cylindrical oven. The dried product contains 10% moisture. The dried bark is then packed in gunny bags. The dried bark is called 'Druggists' bark (quinine content 1.8–2%) in trade.

Yield of Bark

During the first two coppicings, a yield of 4000 kg of dry stem bark per hectare is obtained. At the final stage of uprooting the tree, the yield of the bark may be about 6000 kg/ha.

Isolation of Quinine

The alkaloids are extracted from the powdered bark. Quinine is isolated from the total alkaloids of the bark as quinine sulphate. The commercial preparations contain cinchonidine and dihydroquinine also, and the quinine may be purified by recrystallization to constant specific rotation. The crystalline, efflorescent trihydrate, is a white, odourless, intensely bitter, micro-crystalline powder.

Substitutes for Cinchona Bark

The barks of C. lancifolia, C. ouata, Remijia pedunculata and R. purdieana are often used as substitutes for cinchona bark.

Coleus

Plant Profile

Family	:	Lamiaceae
English name	:	Coleus
Indian name	:	Pashan Bhedi (Sanskrit), Patharchur (Hindi), Makandiberu (Kannada)
Species and varieties	:	Coleus barbatus K–8

Distribution	:	India, Africa (tropical east), Arabia, Brazil Egypt, Nepal, Pakistan, Sri Lanka
Drugs	:	Drugs

Coleus barbatus Syn: Coleus forskohlii Brig., belonging to the family Lamiaceae, is one of the most significant potential medicinal crops of the future as its pharmacopeical properties have been discovered only recently. The tuberous roots are found to be a rich source of forskolin (syn. Coleonol) which is being developed as a drug for hypertension, glaucoma, asthma, congestive heart failures and certain types of cancers. The plant is well known throughout the country and is known as Pashan Bhedi in Sanskrit, 'Patharchur' in Hindi 'Garmalu' in Gujarati and 'Maimnul' in Marathi. In Egypt and Africa, the leaves of C. barbatus are used as an expectorant, emmenagogue and diuretic while its foliage is employed in treating intestinal disorders, and it has been used as a condiment in India for a long time. The tuberous roots of this plant, resembling a carrot in shape and brown in colour, are the commercial parts.

Origin and Distribution

The crop has been distributed all over the tropical and sub-tropical regions of India, Pakistan, Sri Lanka, tropical East Africa, Brazil, Egypt, Arabia and Ethiopia. In India, it is found in the subtropical Himalayan regions from Kumaon to Nepal, Bihar and the Deccan Plateau of Southern India. It is cultivated in parts of Rajasthan, Maharasthra, Karnataka and Tamil Nadu in an area of about 2500 ha.

Description of the Plant

Coleus barbatus (2n = 30) is an aromatic perennial plant 0.5 m tall with thick tubers and with very showy bluish to pale lavender coloured flowers, arranged in whorls on a long spike like racemes. The entire plant is aromatic (whether fresh or dried). The leaves and tubers have quite different odours, the latter being reminiscent of, but quite different from, ginger.

Members of the genus Coleus, have square stems, branched, the nodes are often hairy. The leaves are usually pubescent, narrowed into petioles. The flowers, borne on racemes, are perfect, the calyx is fine-toothed and deflexed in the front. The pale-blue corolla is bilabiate, the lower lobes are elongated and concave. The ovary is four-parted and the stigma is bilobed. Coleus flowers are cross-pollinated by means of wind or insects. The species has four didynamus, with declinate stamens where filaments unite as a sheath at the base. The roots are tuberous, fasiculated, up to 20 cm long and 0.5–2.5 cm thick, they are conical, fusiform, straight, orangish and strongly aromatic.

Varieties

A selection from Karnataka, 'k–8,' is reported to give 0.5% forskolin and a higher tuber yield.

Soil

It thrives best in porous and well-drained soils with a pH ranging from 5.5 to 7. It does not require very fertile soils and can be economically be grown on soils with marginal fertility. The red, sandy loam soils of Karnataka are ideal for the cultivation of this crop.

Climate

Coleus is a crop of the tropics and is found growing well on barren hills at an altitude of about 2400 m, under tropical and subtropical conditions. In India, it is grown mainly in Belgaum (the Sahayadri mountain range of Karnataka) and in Gujarat. The climate here is humid with a RH ranging from 83 to 95% and a temperature of 10–25°C. The annual rainfall is 100–160 cm, mainly between June-September. It is also found to perform well in less humid and warmer regions of South India like Coimbatore, where it is grown as an irrigated crop.

Cultivation

Propagation

Coleus can be propagated by seeds as well as by stem-cuttings. However, propagation through seeds is little difficult and slow and should be used only for the breeding of new varieties. Whereas, propagation by cuttings is very easy and economical to raise this crop on a large scale.

Nursery Raising

The viability of the seeds being very poor (8–10%), a sufficient quantity of fresh seeds has to be sown in well-prepared nursery-beds to obtain good germination. Regular care about watering, weeding and plant protection of the nursery should be taken. In about 15 to 20 days, the germination is completed. When the seedlings are 45 days old and have attained about 8–10 cm height, they are ready for transplanting.

Vegetatiue Propagation

Vegetatively, the crop is propagated through terminal cuttings. Normally. 10–12 cm-long cuttings, comprising of 3–4 pairs of leaves, are sown in already prepared nursery-beds and regular care about shading and watering is taken. The cuttings establish well in the nurseries and there is no problem in their rooting. After about a month's time, when the cuttings have produced sufficient roots, they are transplanted to the main field.

Planting

In most areas, the crop is planted during June-July at the onset of the Southwest monsoon. Before planting, the field is ploughed deep soon after the of pre-monsoon showers and brought to a fine tilth. Further, the land is prepared into ridges and furrows at a spacing of 60 cm and the rooted cuttings or seedlings are planted 20 cm apart within the row.

Manures and Fertilizers

Studies conducted at TNAU, Coimbatore, to standardize the nutritional requirement of this crop have shown that it responds well to the application of N, P and K. A combination of 40 kg N, 60 kg P_2O_5 and 50 kg KO_2/ha was found to be optimum for obtaining the maximum fresh (120 t/ha) and dry (3.982 t/ha) tuber yield from this crop. Half the dose of N, the whole P and whole K may be applied as the basal dose followed by the remaining half N, 30 days after planting as top-dressing.

Irrigation

The first irrigation is given immediately after transplanting, if there are no rains. During the first two weeks after planting, the crop is irrigated once in three days and, thereafter, weekly irrigation is enough to obtain good growth and yield.

Weeding

Due to the frequent irrigations during the initial stages, there is a lot of competition from weeds. In order to obtain economic yields, frequent weeding during the early growth period is desirable.

Pests and Diseases

Pests

The leaf-eating caterpillars, mealy bugs and root-knot nematodes are the important pests that attack this crop. These insects can be controlled by spraying the plants and drenching their roots with 0.1% methyl parathion, while nematodes can be controlled by the application of carbofuran granules at the rate of 20 kg/ha.

Diseases

Among the diseases, bacterial wilt is the major one. The spread of wilt can be controlled by spraying and drenching the soil adjoining the affected plants with 0.2% Captan solution immediately after the appearance of the disease and later after a week's interval.

Harvesting and Yield

Flowers, if any, should be nipped off during the growing period to obtain more biomass of roots. The crop is ready for harvest 4½ to 5 months after planting. The plants are loosened, uprooted, the tubers separated, cleaned and sun-dried for the extraction of 'forskolin'.

On an average, a yield of 1500–2000 kg/ha of dry tubers may be obtained. However, if proper cultivation practices are applied, a yield of up to 2500 kg/ha of dry tubers can be easily obtained.

Costus

Plant Profile

Family	:	Zingiberaceae
English name	:	Costus
Indian name	:	Chanda (Sanskrit), Kebu, Keyu (Hindi) Kondige (Kannada), Kostam (Tamil)
Species and varieties	:	Costus speciosus argyraphyllus
Distribution	:	India, Indo-Malayan region, Nepal
Uses	:	Drugs

Diosgenin, widely used as a starting material in the commercial production of steroidal hormones, is chiefly obtained from certain species of Dioscorea of which Dioscorea deltoidea constitutes the main Indian raw material. This source, however, has limitations in ensuring large supplies on a sustained basis due to its restricted distribution in a few localities in the northwest Himalayas and poor response to domestication. It has become necessary to search for an alternative botanical source which could be easily cultivated under a wide range of agroclimatic conditions and provide the industry with raw material at a comparative low price and on a sustained basis. A number of plants were found to contain steroidal sapogenins, but none could satisfy all the criteria of such a sustitute. The report of the presence of diosgenin in the rhizomes of Costus speciosus (Family: Zingiberaceae) raised hopes of developing such a source, as the plant was widely distributed in India. Diosgenin is further synthesized into sex hormones and steroidal drugs which are widely used for their antifertility, anti-anabolic and anabolic, properties in family planning and health programmes all over the world.

Origin and Distribution

Costus speciosus (Keon) SM is among the five species of stout, perennial herbs indigenous to the Indo-Malayan region. Two varieties of C. speciosus, viz. nepalensis found only in Nepal and Arunachal Pradesh and var. argyrophyllus having a wide distribution in India, have been described.

Description of the Plant

The plant has broad, lanceolate, dorsally silky leaves, borne spirally on stout, erect or sometimes ascending stems. The flowers are white, with a large incurved lip. These are clustered in terminal globose heads, characterized by large and shiny brown or red bracts. Both diploid (2x = 18) and tetraploid forms occur in nature. C. speciosus perenniates through underground rhizomes which lie a few centimeters below the soil surface. The rhizome is differentiated into nodes and internodes. The vegetative activity above the ground lasts for about 7–8 months. Stems sprout during the month of April, flowering commences during July and continues till the end of September and the fruits ripen during the middle of November, after which the leaves are shed and the majority of the canes start drying up. The underground portion remains dormant from December to March.

Soil

The plant can be grown on a variety of soils ranging from coastal alluvium to the heavy, brown, forest type. It grows more luxuriantly on alluvial soils which have a sandy to clay loam texture with a pH of 5.7–7.5.

Climate

It can be grown from sea-level to about 1500 m elevation. But the areas situated at elevations between 400 and 600 m above mean sea-level, having a subtropical climate with an annual rainfall ranging between 1000 and 1 500 mm, produce good quality material. A well-marked dry period between monsoon precipitation results in higher diosgenin content in naturally occuring plant populations. Coastal regions and regions with high annual rainfall and humidity throughout the year bear poor quality raw material, in terms of diosgenin content.

Land Preparation

The land is ploughed 2-3 times and the soil is brought to a fine tilth. FYM @ 15 t/ha is applied and mixed well with the soil and furrows are made 50 cm apart.

Cultivation

Propagation

Although the plant can be propagated from seeds, stem cuttings and rhizomes, commercially it is being propagated only through rhizome cuttings. The selection of rhizomes for planting is, however, important. The rhizomes have a number of nipple-shaped buds, most of them being concentrated around the stem scars and the tips. The formation of buds on the rhizomes is poor during April. The cuttings of rhizome pieces for propagation should have at least two viable buds. Rhizome pieces weighing around 40 g should be selected. About 2000 to 2400 kg of fresh rhizomes are required for planting a hectare of land.

Planting

The rhizome pieces are placed at a depth of 8–10 cm taking care to place the eye-buds facing upwards, horizontally, in rows 50 cm apart and covered with soil. The crop is irrigated immediately after planting. Thick-sized pieces sprout slowly only after 40–45 days of planting. This is due to the eye-buds being dormant on these rhizome pieces which take a longer time to develop, especially in the case of the crop planted in April. After 70–75 days about 90–95% sprouting is obtained. Dessication of young sprouts has been observed during the hot summer months, necessitating a liberal water supply during this period. The best period of planting is from the 3rd week of April to the 3rd week of May. Rhizomes planted during June and July complete germination within 60 days, but yields obtained are low as the crop has a shorter growth period in the first year before dormancy sets in. The active vegetative growth period of the crop is from July to mid-September and the maximum turberization takes place between September to November. The effect of planting on ridges versus flat beds indicates that there is no significant difference in yield due to the method of planting, although a slightly higher yield of fresh rhizomes was obtained from the crop planted on ridges.

Manures and Fertilizers

It is a rhizomatous crop and, to compensate the biomass production, heavy manuring is required. In a trial conducted at Vallanikkara Kerala, the tentative optimum dose for obtaining the maximum yield of diosgenin was observed to be 45 kg N, 30 kg P_2O_5 and 30 kg K_2O along with 15 t/ha of FYM. The FYM and a half-dose of P and K are applied in 2 split doses at 20 and 60 days from the time of planting, the remaining half-dose of P and K is given along with the second dose of N after the 60th day of planting.

Irrigation

The crop requires a liberal supply of water for its successful growth. The crop planted during April and May requires irrigation at least two to three times a month till the outbreak of the monsoon. As September to November is the period of maximum turberization, at least two irrigations should be given in this period. During the dormancy period (December to March), if winter rains are scanty, there is a need for irrigation. If moisture conditions are adequate it has been observed that the crop sprouts towards the end of March' or early April. With a liberal supply of irrigation during the hot summer months, an excellent crop growth is observed. Thus, for a crop to be harvested after a period of 17 to 18 months, 14 to 17 irrigations are required.

Weeding and Interculture

One weeding during the sprouting period of the crop followed by two more, keeps the crop fairly free of weeds. During the period of active vegetative growth (July to September), most of the weeds are suppressed. If the monsoon is erratic, at least one more weeding is required. Weeding once or twice during the dormant season helps in better sprouting of the crop during the next season. During the second year of the crop, the inter-row space between the plants is covered up by the canes and only two to three weedings are necessary before the crop is harvested in August or September.

Pests and Diseases

There are no major pests which affect this crop. However, the following are some diseases which are observed to affect the crop.

1. Rhizome Rot

It is caused by Phytophthora solani and develops very quickly from July to August, in the rhizomes kept in storage after the harvest. Initially, the symptoms start from the tip of the injured portion of the rhizomes. Subsequently, the rhizome becomes light-brown and gives off an offensive odour. Properly harvested rhizomes without injuries can be stored for a long time. No chemical control has been worked out.

2. Pythium Rhizome Rot

The disease is caused by Pythium spinosum. In this disease, the infection in the plant starts from the injured portion of the rhizomes and in later stages the rhizomes turn dirty brown in colour, the leaves become yellowish-brown and finally dry off. The prudent selection of rhizomes from the healthy crop and dipping the rhizomes in fungicidal solutions like Benlate, Bavistin or Dithane-Z-78 has been recommended to control the disease.

3. Leaf Blight

It is caused by Curvularia paradisii. It is very severe from July to September. The symptoms develop on the leaf laminae as small, spherical to irregular spots, light-brown in colour, and in the advanced stages they become dark-brown. The upward curling of young leaves is very common. In the advanced stages, all the leaves fall off, leaving bare stems in the field. This pathogen can be effectively controlled by spraying 0.3% Dithane M–45 at fortnightly intervals.

Harvesting and Yield

It is observed that the diosgenin content is maximum when the crop is in an active stage of vegetative growth, and when it is about 16–17 months old. Two operations, viz. harvesting the aerial shoots and digging out the rhizomes, are required to complete the harvesting in costus. After harvesting the aerial shoots, the most satisfactory and economical method to dig out the rhizomes is to run the tractor drawn cultivator cross-wise twice or thrice over the field and simultaneously collect the uprooted rhizomes manually.

A negligible amount of damage is caused to the rhizomes during this operation. The harvested rhizomes are spread out for a few days and the mud adhering to the rhizomes is removed or, alternatively, the rhizomes are washed under a jet of water from the tubewell discharge pipe. The rhizomes have to be chopped and dried before the diosgenin can be extracted. Hand-chopping is time-consuming and costly, so chopping by mechanical means is employed. Several methods have been tried, but the ordinary wheat thresher with a little modification is found satisfactory for chopping the fresh rhizomes. Very finely chopped material dries quickly (2–3 days). A yield of 28 to 30 t/ha of fresh rhizomes may be obtained in the 1st season's crop (8-9 months) harvested in the dormant season, but the diosgenin yield will be poor on account of low recovery. The crop should, therefore, be harvested after 17-19 months of planting for a better yield of fresh rhizomes and diosgenin content. From the second season's crop (17-19 months), there is a yield of 50 tonnes of fresh rhizomes from which 75–125 kg of crude diosgenin may be obtained.

Cowhedge

Plant Profile

Family	:	Leguminoceae
English name	:	Common Cow hedge
Indian name	:	Atmagupta, Kapikacchuh (Sanskrit) Gonca, Kaunc, Kivacc (Hindi) Turachi Avare, Nasugunnikayi (Kannada) Punaikkali, Punaippidukkan (Tamil)

Species and varieties	:	Mucuna pruriens M. pachylobia M. utilis
Distribution	:	India, Andaman and Nicobar Islands Sri Lanka
Uses	:	Drugs

Mucuna pruriens (L) DC (= Syn. M. prurita Hook), commonly called cowhedge, is an annual, herbaceous climber belonging to the family Leguminoceae. The black-seeded variety of this plant contains L-DOPA (L-3,4 digydroxy phenylalanine) which is a non-protein amino acid and the alkaloids mucunine and mucunadine. The seeds also contain glutathione, mulhingallic acid, a number of alkaloids including nicotine, prurienine, prurienidine and five bases designated as P, Q, R, S and Y. The seed-kernel oil contains sitosterol and lecithin. Biotransformation of L-tyrosine to DOPA has been shown in suspension cultures. Besides, 5-hydroxyptamine and a proteolytic enzyme-mucanin are present. This crop is gaining commercial importance because L-DOPA at 60 glday in 3–4 divided doses is an anti-Parkinsonian and hypertensive drug. The percentage of this chemical on a crude-weight basis is about 4% and in crystalline form 2 to 2.5%. In addition, the seeds have fats, sterols, alkaloids and other amino acids. The plant is said to have antidiabetic properties also. The seeds are astringent, nutritive and have shown hypoglycaemic activity in albino rats. They are also used as an aphrodisiac, as a nervine tonic, in scorpion stings, for leucorrhoea, spermatorrhoea and menstrual disorders. Generally, the seeds are given as a powder in doses of 30–40 g or in the form of a decoction. The roots are pungent, stimulant, diuretic, purgative and tonic. They are recommended for diseases of the nervous system such as facial paralysis, *etc.* An ointment made from the root is used for the treatment of elephantiasis. Besides, the roots are prescribed in fever; powdered and made into a paste and applied to the body in dropsy; a strong infusion mixed with honey is given for cholera also.

In India, 14 species are found in the foothills of the Himalayas, the plains of West Bengal, Madhya Pradesh, Karnataka, Kerala, Andhra Pradesh, Uttar Pradesh, the Andaman and Nicobar Islands and Sri Lanka. Amongst these, M. pachylobia and M. utilis are cultivated for their young leaves and pods which are used as a fodder and a vegetable, respectively.

Description of the Plant

Mucuna is an annual climber. The leaves are pubescent, 3-foliate. Its flowers are 2.5–3.7 em long, purple coloured and are borne in racemes. The fruit is a pod, 10–12 cm long, covered with brown to grey stinging hairs. The seeds are 5–6 mm in size, small, white, black or mottled and number 4–6 in each pod.

Soil

The crop can be grown on a wide range of soils, but well-drained sandy to clayey loam soils are best suited.

Climate

This plant can withstand drought and can be grown in tropical and subtropical climates. Under favourable conditions, the plant grows vigorously. It is sensitive to frost and should not be grown in frost-prone areas.

Season

The crop can be grown both under rain fed and irrigated conditions. The rainfed crop· is sown in June with the onset of the monsoon and the final harvesting is done in December. As an irrigated crop, it can be grown throughout the year. But the best season for sowing the crop is August and the harvesting of this crop is usually done in January-February.

Land Preparation

The field is ploughed well 2–3 times and brought to a fine tilth. About 10 t/ha of FYM is applied to the soil at the time of the last ploughing. Finally, ridges are made along the contours with a spacing of 60 cm between the ridges.

Cultivation

Propagation

The plant is propagated by the direct sowing of seeds. About 50 kg of seeds are required for sowing one hectare of land.

Sowing

The seeds are sown on the inner sides of the ridges (opened at every 60 cm) 60 cm apart, provided that support is given to the plants. If support is not given to individual plants, the seeds can be sown 45 cm apart in 60 cm rows. A basal irrigation may be given before sowing. The seeds germinate in about 7–10 days after sowing.

Fertilizer Application

The crop should be supplied with 100 kg N, 80 kg P and 40 kg K/ha. Phosphorus is applied as a basal dose along with FYM. The entire quantities of N and K are applied in two equal, split doses when the crop is 30 and 60 days old.

Staking of Plants

Mucuna is a twining annual, hence, artificial supports are provided to the vines to get the best yields. Each vine can be supported with sticks 150–180 cm high or tied to an overhead trellis (similar to the type used in Dioscorea) by gunny twines. This method, however, is expensive. The choice of material for the support depends upon the materials available and the cost. Sticks or parts of shrubs will also serve the purpose.

Irrigation

For an irrigated crop, before sowing, a basal irrigation should be given. A second irrigation should be given after the germination of the seeds. Further, the crop is to be irrigated after each top-dressing of fertilizers. The crop needs about 4–6 irrigations from November to December at intervals of 20–25 days. Sometimes, maize or jowar are grown as an intercrop.

Pests and diseases

No serious incidence of pests or diseases has been reported in this crop, except towards the end of the season. At that time, occasionally, leaf eating insects and Cercospora leaf spot are noticed. The crop can be protected from these insects and diseases by two sprays of Malathion at 2 ml along with Dithane M–45 at 1.5 g/l at 30 days of growth and another at 90 days of growth.

Harvesting and Yield

The plants start flowering after about 45 days of growth. Flowering in this crop is continuous and the pods mature periodically. Hence, the dry pods are harvested 3 to 4 times in the season. The pods are sun-dried after harvest. Due to the sun-drying, the pods will break open. Then, the seeds are separated from the husk by winnowing.

Under rain fed conditions, if the crop is left on the ground without staking the plants, a seed yield of 1 500–1 750 kg/ha can be expected. Whereas, under the same conditions, if staking is provided to the crop a seed yield of 3000–3750 kg/ha can be obtained. In irrigated crops with proper staking, the seed yield will be around 5000 kg/ha.

Chemical Evaluation

Aqueous solutions of sodium nitrite (10% W/V), sulphuric acid (10% W/V), sodium hydroxide (20% WN) and hydrochloric acid (1M) are prepared. A standard solution of methyl LDOPA is prepared in 1M hydrochloric acid to obtain the final concentration of 200 mg/ml.

A powdered sample equivalent to 0.2 g of methyl LDOPA is dissolved in 100 ml of 1M hydrochloric acid and filtered after discarding the first 10 ml. Further, 10 ml of the filtrate is diluted to 100 ml with 1M hydrochloric acid.

Now, 5 ml each of sample and standard solutions as well as 5 ml of 1M hydrochloric acid are taken in three separate 50 ml volumetric flasks. To each one of them, 5 ml of sodium nitrite solution is added and the flasks are shaken for 2 minutes. After an addition of 5 ml of sulphuric acid solution, the resultant solution is again kept for 2 minutes and then 6 ml of sodium hydroxide solution is poured into each one of them. The volumes are made-up to the mark with water. The absorbance of the solutions containing the standard and the sample are measured at 490 nm, using the solution containing 1M hydrochloric acid as a blank.

Datura

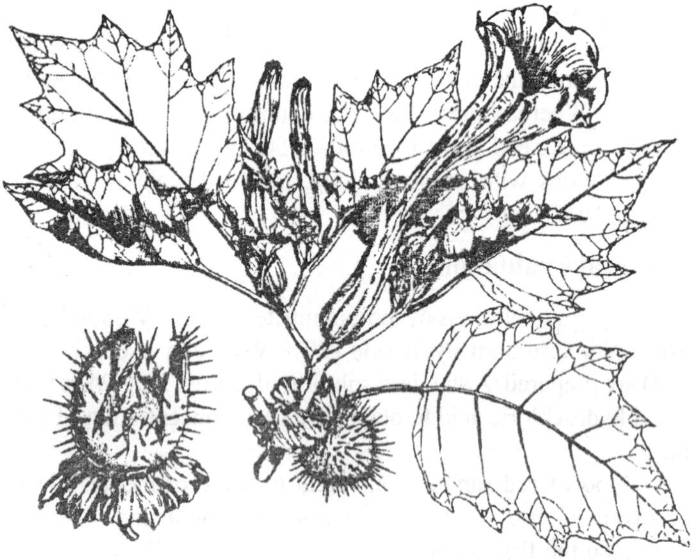

Plant Profile

Family	:	Solanaceae
English name	:	Thorn apple
Indian name	:	Dhatturah, Dhusturah (Sanskrit) Dhattura, Kaladhattura (Hindi) Ummatti (Kannada), Ummattai (Tamil)
Species and varieties	:	Datura stramonium D. innoxia D. metel RRL-Purple RRL-Green

Distribution	:	India, Africa, Asia, Europe, Mexico South America, USA
Uses	:	Drugs

The genus Datura, belonging to the family Solanaceae, consists, of annual or perennial herbs, shrubs and trees growing mostly in the warmer parts of the world. Out of the 15 species reported from the world, 10 species are found in India. Some of them, like Datura stramonium and D. innoxia, are weeds. They are narcotic, poisonous when used in large doses, and medicinal in small doses. Some of the Datura species contain alkaloids, rich in scopolamine, hyoscyamine and metaloidin. Out of the 10 species reported in India, 3 species, namely, D. stramonium, D. innoxia and D. metel, are important drug plants.

Datura Stramonium L. Syn. D.tatala L. (Jimson weed) Stink Weed, Mad Apple, Thorn Apple, Stramonium

Origin and Distribution

It is believed to be a native of South America and is found growing on the hills throughout India up to an altitude of 2400 m. It is a common weed in many parts of the world. This crop is commercially cultivated in the USA and Europe with a view to obtain a drug of uniform potency. In India, it is cultivated in the hills of Uttar Pradesh, Himachal Pradesh, the Kashmir Valley and some parts of Karnataka.

Description of the Species

It is a glabrous or farinose annual, usually 90 cm high, but attaining a height of 180 cm or more in rich soils. The stem is erect with spreading branches. The leaves are pale-green, ovate or triangular-ovate, 12.5–15 cm long and irregularly toothed. The flowers are large, 7.5 to 20 cm long, white or violet, the capsule is erect, 7 × 3.5 cm, ovoid, thickly covered with sharp spines and dehiscing into 4 valves. The seeds are numerous and reniform. Its chromosome number is 2n = 12 and it is a diploid.

Parts of Economic Importance

The leaves and flowering tops are important. The roots, fruits and seeds also contain alkaloids.

Uses

Stramonium is obtained from the dried leaves and flowering tops of D. stramonium. It has a characteristic disagreeable odour and a bitler, unpleasant taste. Stramonium contains 0.3 to 0.5% of tropane alkaloids, chiefly hyoscyamine and small quantities of atropine and scopolamine (hyoscine).

Stramonium is a narcotic, an antispasmodic and anodyne and is used chiefly to relieve the spasms of bronchitis or asthma. The leaves may be made into cigarettes or smoked in a pipe, with or without tobacco, to relieve asthma. They are also used in the treatment of Parkinson's disease.

The alkaloid content of different parts of D. stramonium is as follows.

Leaves	Stem	Roots	Fruits	Seeds
0.41– 0.45%	0.25 – 0.26%	0.21%	0.46%	0.19%

Datura innoxia. Mill. Syn. D. metel avett (2n = 24)

Origin and Distribution

It is a native of Mexico, and is also found growing in the Western Himalayas, the hilly regions of the Western parts of the Deccan Peninsula, and a few other places in India. D. innoxia and D. metel are found mainly distributed in the states of Punjab, Haryana, Rajasthan, Gujarat, Andhra Pradesh, Maharastra, Karnataka and Tamil Nadu.

Description of Species

It is a coarse, bushy annual attaining a height of 90-120 cm. The leaves are dark-green, ovate, often somewhat cardate, about 12.5 cm long and 7.5 cm broad. The flowers are white and fragrant; about 7.5 cm long. The fruits are ovate-conical, about 5.0 cm long and 3.75 cm in diameter, opening at the top into 4 valve-like forms, exposing a long central column bearing numerous light-brown seeds and emits a heavy, narcotic odour.

Parts of Economic Importance

All plant parts, i.e., leaves, flowering tops, fruits and seeds, contain the alkaloid.

Uses

It is used in India for the same purposes as D. stramonium. It is of interest as a possible source of the alkaloid. scopolamine, which is used as a pre-anaesthetic in surgery and childbirth, in opthalmology, and for the prevention of motion sickness in the form of its salt, hyoscine hydrobromide. The alkaloid content of the different parts of this species is as follows:

Leaves	Stem	Roots	Fruits	Seeds
0.25%	0.3%	0.39%	0.12%	0.23–0.25% (0.2%–0.35%)

Datura metel L.Syn. D. fasterosa L. D. alba Nus, D. fasterosa var. alba Nus, Devils trumpet, Black datura

Origin and Distribution

It is a native of Asia or Africa. It occurs throughout India and is occasionally grown in gardens.

Description of the Species

It is a subglabrous, spreading herb sometimes becoming shrubby. The leaves are triangular-ovate in outline and unequal at the base. The flowers are 17.5 cm long, often double or triple, white, violet, reddish-purple or purple on the outside and white within. The fruits are globose, tuberculate or muricate; borne on a short thick peduncle, the capsule dehisces irregularly exposing a mass of closely-packed, light-brown, flat seeds, which nearly fill the interior.

Parts of Economic Importance

All the part contain the alkaloid.

Uses

Used for the same purposes as D. innoxia and D. stramonium.

The aklaloid content of the different parts of this species is as follows:

Leaves and stem	Roots	Fruits
0.1%	0.10%	0.20%

Varieties

Two improved varieties of D. metel have been evolved by RRL, Jammu. They are:

(a) PRL-Purple

This is a early flowering type and the entire plant is purple with dark-green leaves. The yield is about 240–290 q of green herb and 18–24 q/ha of seeds. The alkaloid content varies from 0.24 to 0.36% in the leaves and 0.12 to 0.19% in the seeds.

(b) RRL-Green

A late maturing type, the entire plant is green with light-green leaves. The yield is 210 q of green herb and 1 500-2 000 q/ha of seeds. The alkaloid content varies from 0.24 to 0.28% in the leaves and 0.098 to 0.128% in the seeds.

Soil

Daturas can be grown on a variety of soils but prefer rich alkaline or neutral, clay, loamy soil or those tending to saline-alkaline reaction, rich in organic matter, for their successful growth and yield. Clayey, acidic and/or water-logged soils do not suit this crop.

Climate

Datura is grown in the warmer parts of the world. It can be grown up to an altitude of 2400 m. The plant is sensitive to frost and sheltered situations. Open areas with bright sunshine are therefore preferred for their cultivation.

In the plains, the plant cannot withstand high rainfall or high temperature. Locations with an annual rainfall of 100 cm and a temperature range between 10–15°C in winter and 27–28°C in May-June are ideal for its cultivation.

Season of Sowing

1. D. innoxio is grown as a winter or spring crop and the seeds are Sown in October or in late January, respectively.
2. D. metel is grown as summer crop and the seeds are sown in March.

3. D. stramonium is grown in temperate areas like the Kashmir Valley. The seeds are sown in March-April and in November in the North Indian plains. The seedlings are transplanted to the field in May-June in temperate areas and in November-December in subtropical areas.

Land Preparation

The land is ploughed 2–3 times followed by planking. Weeds and stubble are removed and FYM with a basal dose of fertilizers is incorporated into the soil during the land preparation.

Cultivation of Datura Species

Propagation

Daturas can be grown directly through seeds or by transplanting nursery-grown seedlings.

Nursery Raising

The nursery-beds are prepared by mixing well-rotted FYM into the beds. The seeds are sown by the broadcast method on the beds and covered with a layer of sand and FYM. The beds are kept moist by spraying water. About 2 kg of seeds are required to raise seedlings for planting one hectare.

Problem of Seed Germination

The germination of datura seeds is low and irregular due to the presence of an inhibitor. Seed germination can be enhanced by soaking the seeds overnight in water and washing them repeatedly 2 to 3 times with fresh water before sowing. The seed germination can also be hastened by alternate exposure to freezing and thawing This treatment makes the seed coat weaker.

Direct Sowing

The seeds are sown in rows 45-60 cm apart and covered with soil. About 7–8 kg of seeds are sufficient for sowing one hectare. The seeds start germinating within a fortnight and in a month's time the germination is complete. Weeding and thinning is done when plants are 10-12 cm high, keeping a plant-to-plant distance of 30–45 cm within the rows.

Transplanting

In the case of the crops raised by transplanting, the seedlings are transplanted when they are 8–12 cm tall and possess four leaves, at a distance of 30–45 cm in rows which are 45–60 cm apart. A spacing of 75 × 75 cm is recommended for obtaining higher yield and alkaloid content of seeds, for the crop planted in December and January.

Manures and fertilizers

Daturas respond very well to the application of manures and fertilizers. In a study conducted at the University of Agricultural Sciences, Bangalore, it was found that the application of 10 tonnes of FYM along with 60–90 kg of N, 3MO kg of P_2O_5 and 40–50 kg of K_2O are optimum for obtaining higher herbage and alkaloid yield in D. stramonium.

A half-dose of nitrogen and the full dose of phosphorus and potash are applied at transplanting, and the remaing half-dose of nitrogen can be given after two months of transplanting. For D. innoxia, the application of FYM @ 2t/ha before sowingl planting, 80 kg Nand 90 kg P_2O_5/ha during the vegetative growth is recommended. Phosphorus applied at the time of sowing is also reported to give better yields. The alkaloid content is reported to be affected by micronutrients.

Irrigation

The first irrigation is arranged immediately after the sowing or transplanting, if there are no rains. The subsequent irrigations may be given at intervals of 8–10 days depending upon the weather conditions, till the final harvest.

Weeding and Hoeing

The first weeding and hoeing are done when the plants attain 1012 cm height, *i.e.*, about one month after transplanting, and the second hoeing and weeding are done after two months.

Chemical Weed Control

Dimid (diphenamid) @ 5 kg/ha applied a few days before the

crop emergence, provides effective weed control and improves crop growth and yield.

Pests and Diseases

No major pests have been reported on this crop in India. However, thrips act as vectors in transmitting the mosaic virus, which can be controlled by spraying a suitable insecticide like Metasystox (1.5 ml/lit.).

Diseases

In India, the plant is very hardy and is not affected by any serious disease. However, some of the important diseases observed elsewhere are:

(i) Wilt

This disease is caused by Sclerotium rolfsii. Infested plants show initial symptoms of drooping leaves and finally wilting of the entire plant. The characteristic browning and rotting has been observed in the roots and the collar region of the infested plants. No control measure is suggested for this disease. However, field sanitation and crop rotation are recommneded.

(ii) Root and foot-rot

The causal organism for this disease is Corticium solani. Seedlings show symptoms of damping-off and the plants are killed within few days in the nursery. In mature plants, the leaves become yellow and there is drooping of apical parts. The infested portion at the soil surface becomes dirty brown in colour and the tissues get separated from the bark. Young seedlings can be saved by drenching them with a solution of copper funigicide (0.3%) in the nursery.

(iii) Leaf spot

This disease is caused by Alternaria tennuissima in D. metel and D. innoxia. The disease begins with the characteristic dark-brown spots of various sizes, which are round to oval and sometimes irregular in shape. The spots gradually enlarge with concentric rings and become necrotic at later stages, which leads to the withering and drooping of the leaves. No control measure has been worked out for this disease, on this crop.

(iv) Viral diseases

Several viral diseases like Distortion mosaic, Enation mosaic, Rugose leaf curl, Little leaf and Mosaic have been reported under this group of crops. Some of the control measures recommended to control them are: uprooting the affected plants from the nursery as well as from the main field, the use of seeds from disease-free fields and control of vectors like thrips (mosaic virus) and plant hopper (rugose leaf curl) by spraying any suitable insecticide.

Harvesting and Yield

(1) D. stramonium

The crop is ready for harvest 6-7 months after sowing. While harvesting, the entire plant is cut when the fruits are mature but green, and partially dried in the sun or in the shade. The leaves are stripped and separately dried. The seeds are shaken off from the capsules when the fruits begin to burst. An output of 11.2–17 q/ha of leaves and 8 q/ha of seeds may be expected.

(2) D. innoxia

In India, this species is grown for its seeds. The crop is ready 6–8 months after sowing for the final harvest. Plantations raised in winter (October) and spring (late January) start flowering by the middle of February and by the beginning of April, respectively (5 months after sowing). The time required for the fruits to mature is one month after flowering. Fully-grown fruits, when still green, are plucked by hand 2–3 times before the final harvest which is done by cutting the whole plant with a sickle from the base, by the end of May in the former case and by the end of June in the latter case (approximately 8 months after sowing in case of the winter-sown crop and 6 months in the case of the spring-sown crop).

The harvested fruits are spread out under the sun until they open. The open fruits are threshed with a stick to separate the seeds, which are dried in the sun and packed in gunny bags when fully dried. A yield of 1200–1700 kg/ha of dried seeds is obtained, with an alkaloid content varying from 0.2 to 0.35%.

(3) D. metel

These plants attain optimum vegetative growth as well as the highest percentage of alkaloids by July (5 months after sowing), when the first harvest is taken. All the tender branches along with the leaves are cut. The herbage is dried in the shade immediately.

The harvested plants regenerate and two more harvests are taken, with a second harvest in late August and a third harvest in October. In the last harvest plants can be cut from the base, if a ratoon crop is not desired. Besides foliage, a sizeable number of fruits are also harvested as and when they ripen. A yield of 240–290 q/ha of green herb and 18–24 q/ha of seeds is obtained from the variety RRL purple, and a yield of 210 q/ha of green herb and 15–20 q/ha of seeds can be obtained from the variety, RRL-Green.

Digitalis

Plant Profile

Family	:	Serophulariaceae
English name	:	Foxglove
Indian name	:	Titpushpi (Sanskrit)
Species and varieties	:	Datura purpurea L D. lanata Ehrh.
Distribution	:	India, the British Islands, Canada Western Europe,

		France, Germany, Hungary Mexico, UK, USA and some Asian countries
Uses	:	Drugs

Digitalis, commonly known as Foxglove, belong, to the family Scrophulariaceae. There are many species under the genus Digitalis of which D. purpurea L. and D. lanata Ehrh. are the most important for their glycoside content, used in life-saving medicines. The leaves of the recommended varieties are a rich source of digoxin, an active cardiac glycoside. Besides, the leaves also contain digitoxin and gitoxin. Other important glycosides are digitoxigenin, glucodigit-oxigenin-bis-digitotoxoside, glucogitaloxigein-bisdigitoxoside, glucovatromonoside, glucogitoroside, glucolanadoxin, varadoxin, stropeside and anthraquinone. Digitalis is used mainly for its effect on the cardiovascular system, *i.e.* increasing the force of systolic contractions and the efficiency of a decompensated heart. It is used as a myocardial stimulant in congested heart failure, auricular flutter and rapid auricular fibrillation. In the case of burns, it is more selective in preserving the cells severely injured by heat.

Presently, the major portion of the raw material required by leading pharmaceutical industries in India, is being met by imports. The price of digitoxin amounts to crores of rupees. In view of the importance of the crop for its glycoside content, improved cultivation practices have been developed through a series of experiments conducted at Solan and Kodaikanal.

Origin and Distribution

Digitalis is a native of Western Europe, the British Islands and the erstwhile USSR countries. The crop is being cultivated in France, Germany, UK, Hungary, USA and other Asian countries over centuries for medicinal purposes. It has been introduced as an ornamental plant throughout North America, Canada, Mexico, Central America and Asia.

The northern states like Jammu and Kashmir, Himachal Pradesh and the hilly and elevated areas in South India have vast potential to produce a sufficient quantity of leaves and seeds to meet the demands of the pharmaceutical industries in the country.

Description of the Plant

Digitalis (2n = 56) is a family of herbaceous, erect, branched biennials attaining a height of 60–90 cm in nature. The leaves are simple, exstipulate, alternate, opposite or whorled. The flowers are zygomorphic, bilipped, pentamerous, hypogynous and bisexual with 4–5 partite bracts. The sepals are gamosepalous and lobed, there are 4 petals which are gamopetalous, bilipped, sometimes spurred and with Imbricate aestivation. There are 2–4 stamens. The anthers are bithecate, hairy, dehiscing by longitudinal slits and epipetalous. The ovary is bicarpellary, syncarpous, superior, with many ovules and a single or bilobed style. The stigma is bilobed and the fruit is a capsule. Digitalis is a insect-pollinated plant, bees being the main pollinators.

Varieties

The strain E.C. 115996 has been reported to give a higher foliage yield with high glycoside content. This parent material has been used in crossing and selection amongst progenies to identify a few promising selections, namely, 0–76, 0–21, OYF and OPF, which have consistently given higher yields over the years.

Soil

Foxglove requires a well-drained, sandy loam soil, rich in organic matter. Ill-drained and water-logged conditions are detrimental to the crop. Clay and sandy soils are also harmful. While lime-rich soils are essential for proper growth and glycoside content in the leaves of D. lanata, it is harmful for D. purpurea which requires acidic soils.

Climate

The crop prefers a warm climate. The two important species, viz. D. purpurea and D. lanata, that are suitable and promoted for commercial cultivation in India, perform best in a temperature ranging between 20°C and 30°C. The growth and development of the plants as well as the content of their active principles is maximum; furthermore, the seed germination is best in this temperature range. For the luxuriant growth of the crop, well-distributed rainfall and good sunshine during winter are ideal. It is found growing successfully up to an altitude of 200 m above mean sea-level (MSL).

Season

The time of planting is very critical for this crop. The best planting period is mid-April in cooler regions like Solan (Himachal Pradesh) and the second fortnight of July at Kodaikanal (Tamil Nadu). Planting in August in the other hilly regions with a higher elevation in South India, registered a higher yield of leaves and glycoside content.

Land Preparation

As the crop is a biennial, the field preparation should be thorough. Plough the land 4 to 5 times using a disk and harrow, remove the weeds and bring the soil to a fine tilth. Incorporate the FYM during the final ploughing for thorough mixing with the soil. Divide the field into plots of convenient size (2 to 3 sq m) for distributing the water evenly.

Cultivation

Propagation

Digitalis is propagated from seeds, either by direct seeding (linesowing/broadcasting), or by first raising the seedlings in a nursery and then transplanting them.

Seed Material

Seeds produced in the Solan and Chamba districts (H.P) are the best for the successful cultivation of this crop in India. About 1 500 seeds weigh one gram. Fresh seeds should be used for sowing, since the seeds lose viability after two years of harvesting. The seed should be presprouted to hasten germination by soaking them in water and incubating them at about 30°C for 2 to 3 days. About 5-8 kg of seed as required for directly seeding one hectare, while for nursery raising Just 1 to 2 kg of seeds will suffice.

Direct Sowing

The field, thus prepared, can be used for direct seeding by linesowing at a depth of not more than 2 cm to avoid failure of seed germination. A spacing of 45 cm between rows and 30 cm between plants in a row are maintained in both methods of planting.

Nursery Raising and Transplanting

Raised nursery-beds (10–12 numbers) of the dimension: 3 m long, 1m wide and 10 cm high are prepared. The soil for the nursery beds is ennched by organic matter. To each bed, 3 to 4 basketfuls of well composed FYM is incorporated and the land is brought to a fine tilth. The seeds are sown in lines 4–6 cm apart· the surface of the bed is covered with a 1cm layer of sand. The beds are irrigated lightly every evening with rose-fitted cans. The seeds germinate in 15–20 days and the seedlings will be ready for transplanting about 35 to 45 days after sowing, when they have attained a height of 8–10 cm.

The field is kept ready by preparing ridges and furrows at a 45–50 cm distance. When the seedlings attain a height of 8–10 cm to (35–45 days after sowing), they are transplanted to the main field at a spacing of 30 cm apart, in rows.

Manures and Fertilizers

FYM is applied at the time of land preparation while fertilizers are applied in the furrows and incorporated in the soil thoroughly immediately after transplanting. The application of 40-60 tonnes of well-decomposed FYM along with 100–150 kg N, 50 kg P_2O_5 and 25–30 kg K_2O is the optimum. One-fourth of the nitrogen, supplemented in the form of calcium ammonium nitrate, along with a full dose of P and K should be applied at the time of planting. The remaining N is applied in 3 split doses, i.e. first top-dressing at 60 days after transplanting, the second and third dose is applied at 1% to 2 month intervals thereafter. The application of N, P and K is repeated following the above schedule during the second year also. The glycoside content has been found to increase by application of trace elements like molybdenum and magnesium.

Irrigation

During summer (April, June and July), the crop is watered at an interval of 7 to 10 days, then 2 to 3 irrigations will be sufficient each month until the dormant period which sets in by November. The first harvest is done just before that. Then one or two irrigations a month will be required till the new leaves are formed. Weekly irrigations are recommended till the leaves mature. However, during the mon-soon, watering can be suspended as there will be sufficient precipitation to take care of the crop.

Interculture and Weeding

Weeds, by competing for light, moisture and nutrients and also acting as host for many diseases and pests cause a drastic reduction in leaf yield. Hence, the crop must be kept completely free of weeds by either chemical or manual weeding. During the rainy season, hand weeding is preferred. During summer as well as after each harvest hoeing should be attended to. At least two hoeings before the first harvest and another afterwards are important.

Pests and Diseases

The occurrence of insect pests is not serious. Leaf blight caused by Alternaria spp. is a common disease of foxglove. Brown spots appear on the leaves, which enlarge, covering the entire leaf surface and killing the foilage in a few days. Spraying any copper fungicide at the rate of 0.1 to 0.2% effectively controls this disease.

Occasionally, there is an occurrence of curly top, a mosaic and a broad-bean wilt virus. In addition to the chemical control, destroy the alternate hosts and control the vectors in the vicinity. The affected digitalis plants should be pulled out and destroyed.

The other diseases of this crop reported from Europe are anthracnose caused by the fungus Collitotrichum Juscus, two types of leaf spots: one caused by Septoria digitalis and the other by Phyllosticta digitalis, and crown rot caused by Botrytis cineria.

Harvesting of Leaves

While harvesting leaves, attention has to be paid to aspects like the age of the leaves, the time of the day and the length of the leaves. These points are important with respect to the glycoside content. The cardio-activity of the leaves is higher in species like D. purpurea and D. lanata, when the leaves are harvested during the hotter part of the day. The glycoside content in the leaves is highest around midday and the lowest at midnight, since the glycoside quickly disintegrates in the dark. The leaves are well developed when they attain 8-10 cm length and are ready for the first harvest about 5 months after planting. The subsequent harvests will be at a 1½ to 2 month's interval then, afterwards, during the first year of the crop. The plants undergo dormancy after the last harvest. In the second year, the crop can give two harvests, besides seed production. The first harvest of leaves in the second year will be in March-April, approximately about 3 to 4 months after the 1st harvest of the previous year. The final harvest will be about 2 months later. Although the crop is biennial and the leaves are potential glucoside yielders during both the years, it is suggested that the crop is grown as an annual for leaf production, unless meant for seed production in which case it can be grown as a biennial.

Processing of Leaves

Drying and storing the leaves are the other important activities. Great care should be exercised to prevent the loss of the active ingredients. In hotter regions, the drying process is easy and economical, as natural solar energy can be utilized for drying by spreading the leaves in layers on clean yards and turning them over occasionally to obtain a minimum moisture content. This may take 2 to 3 days on hot days.

The alternate method is dehydration, by passing hot air at 50–60°C over the leaves and upturning them. 'This process should, however, be at the appropriate temperature and with the minimum time-lag. Fermentation of leaves during drying should be avoided by turning the leaves occasionally and exposing them to the heat source, otherwise, enzymatic activity leads to the deterioration of glycosides. Higher temperatures, beyond 70°C, will also lead to a loss of active ingredients.

Storing the leaves after drying is yet another factor to take into account for carefully preserving the glycosides. The dried leaves should be stored in airtight containers like polythene bags with some dehydrating substance such as silica gel or activated alumina. The bags should be kept in a cool and dry place.

Yield of Leaves and Glycoside

The variety, EC 115996, gives a higher glycoside content of 0.651.20% It gives a higher foliage yield of 15 t/ha on a fresh-weight basis and about 3 tonnes on a dry-weight basis.

Estimation of Total Glycosides in Leaves

For this purpose, 5 g of powdered, dry sample of Digitalis leaves is taken and added to 80 ml of alcohol (20%), shaken for 6 hours and filtered. The residue is again extracted two times, each time using 80 ml of alcohol (20%) and with chloroform (20 ml × 5) in a separating funnel. All the chloroform extracts (lower layers) are combined, transferred to a conical flask and dried over anhydrous Na_2SO_4.The chloroform layer is decanted into a RB. flask, Na_2SO_4 is washed with chloroform (15 ml × 3) and the washings are added to the main chloroform bulk and the chloroform is evaporated to dryness. The dried residue is dissolved in a mixture of $CHCl_3$: MeOH (@ 1:1) and filtered, the filtrate is evaporated to dryness over a water-bath on a weighed watch-glass and the residue weighed as the total glycoside content.

Dioscorea

Plant Profile

Family	:	Diascoreaceae
English name	:-	Greater yam, Asiatic yam
Indian name	:	Alukam (Sanskrit), Chupri alu, Khamalu (Hindi) Perumvallikilanku, Kappan kaccil (Tamil)
Species and varieties	:	Dioscorea composita Hemsl

		D. flon'bunda Mart and Gal D. deltoidea Wall RRL-Green
Distribution	:	India, Africa, Asia, Europe, Mexico, South America, USA
Uses	:	Drugs

Steroidal drugs alone constitute about 6% of the total production of pharmaceuticals.

They are one of the costliest and most important medicines used throughout the world. Because of their antifertility properties, they have gained tremendous importance in the health and family planning programmes of the developing countries. Diosgenin, a steroidal sapogenin obtained from the rhizomes of various species of Dioscorea (Family: Diascoreaceae), is the major base chemical for several steroid hormones including sex hormones, cortisones, other corticosteroids and the active ingredient in the oral contraceptive pill. The other important sapogenins found are yamogenin, botogenin and kryptogenin. Minor sapogenins like pannogenin and tigogenin are also found in certain cases.It is estimated that the world production of diosgenin is 1000 tonnes and of other precursors is 1200 tonnes. Mexico is the largest producer of diosgenin, producing about 750 tonnes annually. The annual diosgenin requirement of India in 1983-84 was estimated to be around 242 tonnes against the 25–30 tonnes produced indigenously.

Botany

Plants of the genus Dioscorea, commonly called medicinal yams, are perennial, climbing herbs with tubers or rhizomes. Approximately 600 species have been reported to occur throughout the world, mostly in tropical and subtropical regions and, to a limited extent, in temperate regions also. Of these, only 15 species are known to contain steroidal sapogenins, chiefly diosgenin.

Out of the various species tested, D. composita Hemsl and D. jloribunda Mart and Gal. found growing wild in Central America and D. delta idea Wall, found in the North-western Himalayas, are the main species from which diosgenin is extracted commercially. All the the three species are dioecious.

Species and Varieties

(1) D. deltoidea Wall (2n = 20)

It is an indigenous species found growing wild in the North-western Himalayas. It produces very slender vines and is very weak. It is propagated by tuber pieces, but the regeneration of tubers is so slow that it takes about 7-10 years to fully develop even in its natural habitat of temperate regions. Hence, the cultivation of this species on a commercial scale is not attractive to farmers.

(2) D. jloribunda Mart. and Gal. (2n = 36)

It is a native of Mexico (Central America) and is grown in Karnataka. Goa, Assam, Meghalaya and the Andaman Islands. This species can be very easily propagated from tuber pieces and seeds. The plant is dioecious in nature, a robust climber twining to the left. The branches are thick and it produces yellow, compact tubers at a shallow depth. The leaves are petiolate and spirally disposed on the stem.

The male flowers are solitary or in groups of 2–3, dark-brown or green and sessile with 6 stamens. The male spike is short and solitary, with 6 staminodes and a hypogymous stigma with a bifid apex (divided into two). The diosgnin content varies from 2–7% depending upon the age of the tubers. The seeds germinate within 3–4 weeks and produce vigorous seedlings that establish well in the field.

Three varieties of D. floribunda have been released for cultivation. The varieties FB(C)-l strain and Arka Upkar were released from the Indian Institute of Horticultural Research (IIHR), Bangalore, while Pusa-l by the Indian Agricultural Research Institute (I CAR), New Delhi. These three arieties adapt well to tropical and subtropical regions. The salient features of these varieties are as follows.

FB (C)-1

This is a composite strain from introduced clonal material from Central America. The plants are vigorous and relatively free from pests and diseases. The tuber yield is 20 000–25 000 kg/yr (1 kg/plant) or 60 000 kg for a 2-year-old crop (2.5 kg/plant). The diosgenin content of the dried roots varies from 2.5 to 3%.

Arka Upkar

A high yielding clonal selection from FB (C)-1 with intense dark-green leaves and a vigorous growth. The tuber yield exceeds 6000 kg (in a 2-year-old crop) and the diosgenin content ranges from 3.5 to 4.0%

Pusa-1

A selection from germplasm with a tuber yield of 1.5 kg/vine after 18 months.

(3) D. composita Hemsl. (2n = 36)

This species is also a native of Mexico and has been domesticated in several tropical countries. It is a robust climber, twining to the left, which produces large, thick leaves. The underground portion comprises of thick, fleshy, branching tubers which are long white in colour and grow deep. This species is mostly propagated from seeds rather than from tubers as the rotting of tubers is pronounced.

Other important silpogenin-bearing species are D. jriedrichsthali R.Kunth. D. spiculiflora Hemsl, and D. prazeri. Prain and Burk.

Soil

Dioscorea can be grown in several types of soils. Light or sandy soils require heavy irrigation and fertilization, whereas heavy clay soils restrict tuber growth and harvest and often create water-logging. The best yields are obtained in medium loam and in deep soils which are rich in organic matter. In red soils, the best growth has been observed for both D. floribunda and D. composita. It tolerates wide variation in soil pH, but highly acidic and highly alkaline soils should be avoided.

Climate

The different species of Dioscorea are found growing in different climates, but while D. floribunda and D. composita are more suited to the tropics, D. deltoida is reported to be a suitable species for temperate locations. It grows well in the temperate regions of Kashmir and Himachal Pradesh.

Preparation of Land and Planting

The field should be ploughed and harrowed several times, levelled properly and drainage channels should be made. Since yams have a high requirement of organic matter for good tuber formation, a sufficient quantity (20–25 t/ha) of FYM is incorporated at the time of land preparation. A spacing of 45×30 cm for a 1-year-old crop and 60×45 cm for a 2-year-old crop in D. floribunda is found to be optimum under irrigated conditions. For planting, deep furrows are made at 60 cm distance and sprouted tubers are planted in furrows at 5 cm depth. After sprouting is completed, earthing up is done, utilizing the soil from the ridges. D. composita and D. deltoidea are reported to give higher yields at spacings of 60 × 30 cm and 30 × 70 cm, respectively.

Cultivation

Propagation

The yams can be propagated either by seeds, rhizome pieces or stem-cuttings. In India, commercial plantations are raised from tubercuttings. Seed progeny is variable and takes a longer time to start yielding tubers, compared to plants raised from tubers. The choice of propagating material will depend on the cost of planting and the prevailing climatic conditions of the region.

(a) Propagation from Tuber Pieces

This crop grows best from tuber pieces. Tubers or rhizomes are divided into approximately 50–60 g pieces for planting. The growth of plants is slow and the yield lower if smaller pieces are used for planting. There are 3 types of pieces: (1) Crowns (stem end), (2) Medians (middle portion) and (3) Tips (distal ends). Crowns produce new shoots within 30 days of planting, while the others take nearly 100 days to sprout. Besides, the crown portion contains less diosgenin compared to the median and the tips, hence the latter can be used for the extraction of alkaloids and the former can be used for propagation. But, if there is a shortage of planting material, the median and tips can also be used for planting.

In order to avoid the rotting of tubers (before sprouting), only healthy tubers should be selected the healthy tubers must then be dipped in benlate fungicide (0.3%) for 5 minutes followed by dusting the cut ends with 0.3% benlate powder before planting or storage. Before planting the tubers directly in the main field, they should be stored in a moist, well-aerated rooting medium until shoot growth commences. The benlate-treated tuber pieces should be kept in raised beds in the shade, covered with sand and watered daily. After 30 days the sand may be removed and the sprouted crowns taken out and planted in the field. The median and tip portions are again covered with sand and watered regularly. Subsequently, after 60–75 days when they have completely sprouted, they can also be planted in the field.

Season of Planting

The tuber pieces can be planted either in February-March or June-July. In Karnataka, February-March planting is better. And for medians and tips, it is better to plant from the middle of January because they take more time to sprout compared to the crowns. The new sprouts will grow vigorously during the rainy season which commences from June.

(b) Propagation by Seeds

Propagation through seeds is much more successful in D. floribunda and D. composita, as compared to the other temperate species. The seed has a wide membranous wing that can be removed without affecting germination. The seeds can be sown either in raised beds in the shade (with a mixture of loamy soils and FYM) or in polythene bags (filled with sand, soil and FYM). The planting depth should not be more than 1.25 cm and frequent watering of the beds is essential. The seeds germinate within three weeks and are ready for transplanting in 3–4 months. The seedlings should be supported immediately.

The best season for transplanting the seedlings to the field is just before the start of rains, *i.e.*, in June in South India, but in North India, this can be done at any time except during the winter.

Seed Production

Since dioscorea is a dioecious plant, female and male plants should be grown close to obtain the seeds. Under South Indian conditions, excellent seed set is noticed. Flowering starts from

August and the seeds mature from December to February. The pods turn brown and dehisce along with the upper margin. The harvested seeds should be stored in airtight plastic bags where they will remain viable for 3–5 years.

(c) Propagation by Stem-cuttings

D. floribunda can be propagated by stem cuttings with 80% success. The vines should be raised from 50-100 g tuber pieces in the greenhouse. One or two month-old vines, are taken and cut into single node cuttings, each with one leaf. They are planted in sand-beds keeping the leaf blade above the sand. Before planting, the cuttings should be treated with 1 ppm 2, 4–D and 0.1 % benlate for 4 hours. The beds should be watered regularly. After rooting, the cuttings are transplanted to polythene bags and produce about ten leaves in a period of two months.

Propagation of D. floribunda can also be done by air- and groundlayering. Application of NAA at 100 ppm was found to be most effective for getting 80–90% rooting.

(d) In Vitro Propagation

Single node leaf cuttings of D. floribunda, when used as explants, give 100% rooting when grown in vitro. In D. deltoidea, both single nodes and apices have been successfully used as explants for propagation. Shoot apices explants are reported to produce fill-fiedged plants after 90–100 days. In vitro-raised single node cuttings were ready for planting in the field after 50-60 days.

Provision of Support

The vines need support for their optimum growth, as this exposes the maximum number of leaves to the sunlight. The system tried successfully at Bangalore consists of 2 m-high, stone pillars spaced 9 m apart in the field. Galvanized wires (Nos. 6 and 8) are used on the boundaries and the interconnecting wires can be of 12 gauge thickness. The vines are supported on coir ropes tied to the wires and pegged to the ground by wooden stakes.

Fertilizer Application

Yams respond very well to fertilizer application. A fertilizer dose of 300 kg N, 150 kg P and 150 kg K/ha, has been found to be optimum for a one-year-old crop of D. floribunda. The entire quantity of P should be applied as a basal dose, while N and K are given in four equal split doses at bimonthly intervals commencing from 2 months after sprouting. For D. deltoidea, for a 4-year-old crop, the recommended fertilizers under Kashmir conditions are 40 kg N, 80 kg P_2O_5 and 60 kg KO_2 in split doses at an interval of one month from the time of planting in the first year and the same quantity in the second year. In the third year, only N at 30 kg/ha has to be applied in two split doses. For increasing the tuber yield and diosgenin content, the application of S, Ca and Mg has also been recommended.

Irrigation

The crop needs irrigation frequently during summer months. An interval of 4 to 5 days in summer and 7 to 10 days in winter is desirable for the proper growth and development of this crop and for economic yields.

Interculture

In the initial stages, yam plantations require more frequent hand weeding than in the later stages. Chemical herbicides have also been tried, but are not recommended commercially. Studies have revealed that it is feasible to intercrop medicinal yam with short duration crops like cowpea, cluster beans and kidney beans, where irrigation is available. This practice not only minimizes the growth of weeds but also gives additional returns. Apart from this, intercropping also helps in reducing the requirement of nitrogen as the crops recommended above are leguminous. Cowpea, cluster beans and french beans are the recommended intercrops for Dioscorea.

Plant Protection

Pests

The two important pests affecting this crop are aphids and red spider mites which can easily be controlled by spraying 0.5–1.0% of Kelthane (1 ml/l) or Rogor (2 ml/l). Cut-worms are also reported to damage this crop. No major disease has been reported to affect this crop. The only disease which affects the crop is the rotting of D.

floribunda tuber pieces during storage in sand-beds. This can be controlled by treating the tubers with 0.3% benlate solution. Leaf spot disease, caused by Drechelera sorokiniana, is also reported to occur in D. composita, resulting in defoliation and death of the seedlings. The disease can be effectively controlled by spraying Benlate (0.1 %). Besides, Cercospora, Remularia and late blight caused by Phytophthora infestans are reported from Tamil Nadu and can be controlled by Dithene 2–78 (0.3%) or any other copper fungicide. Collar rot of seedlings, which is a nursery disease, can be effectively controlled by spraying Brassicol (0.5%).

Duration of the Crop and Yield

The diosgenin content of the tubers tends to increase (within certain limits) with age. Studies have shown that the growing of O. floribunda as a two-year crop is economical. An average yield of 15 to 20 t/ha of fresh tubers can be obtained during the first year, and up to 40 to 50 t/ha during the second year. The diosgenin content of the tubers tends to increase, on an average, from 2.5–3.0% in the first year to 3.0–3.5% in the second year. The cost of cultivation is, however, much less during the second year.

Harvesting and Marketing

Experiments have shown that D. deltoidea should be harvested only after three years to get the optimum yield from the crop with the maximum diosgenin content. Similarly, O. floribunda should be kept at least for two years in the field for viable yields. Generally, the tubers are harvested during February-March. Harvesting can be done by manual labour with pickaxes. The tubers are harvested when the plants are in a dormant condition to obtain the maximum yield of diosgenin. In studies conducted on the effect of gamma radiation on tuber yield and diosgenin content of D. floribunda, it was found that radiation decreases the tuber yields, but the diosgenin content showed a wide range of variation. The diosgenin content in 1, 2 and 3 treatments was 2.6,3.6 and 2.8%, respectively, against 2.9% in the control. A number of drug companies, using diosgenin as raw material, are interested in buying this crop grown on a contract basis. On the basis of the prevailing market conditions with dry tubers, a net profit of about Rs. 30000/ha in a one-year-old crop and Rs. 60000/ha in a two-year-old crop can be obtained.

Duboisia

Plant Profile

Family	:	Solanaceae
English name	:	Corkwood tree
Indian name	:	Bendumara (Kannada)
Species and varieties	:	Duboisia myoporoides D. leichhardtii D. hopwoodii

Distribution	:	India, Australia (Queensland), New Caledonia Japan, northern and central New South Wales
Uses	:	Drugs

Dubosia, belonging to the family Solanaceae, is a subtropical shrub endemic to the Australian region. It is commonly called the 'corkwood tree' and is represented by three species, *viz.*: D. myoporoides, D. leichhardtii and D. hopwoodii. The leaves of the first two species are rich in hyoscine and hyoscyamine, whereas the third one contains tobacco alkaloids, nicotine and nor-nicotine. Recently, the roots of D. hopwoodii have been found to contain substantial amounts of hyoscyamine and hyoscine. Hyoscine, hyoscyamine and atropine are used extensively in medicines throughout the world because of their mydriatic, antichloinergic and antispasmodic action. Hyoscyamine and atropine are used in respiratory diseases like asthma, intestinal disorders, colic pain and peptic ulcers. Hyoscine is used for the treatment of motion sickness and asthma.

The leaves contain a complex of tropane alkaloids including hyoscine (scoplamine), hyoscyamine, nor-hyoscyamine, valeroidine, tigloidine, poroidine, isoporoidine, isopelletierine and anabasine. The constituents vary, not only with the stage of growth, but also with the geographical location. The total alkaloid content of the leaves has been found to vary from 2.75 to 4.90%, of which hyoscyamine was 56.7%, hyoscine 23.3% and other alkaloids 20.0%. On an average, the plant gives a crop containing 3% of total alkaloids.

In the wood of the young root, tropine is the principal alkaloid, with smaller amounts of vatropine, hyoscyamine and valeroidine. The root-bark possesses hyoscine, atropine, tropine and apohyoscine; while the wood of the mature roots contains hyoscyamine, hyoscine, apohyscine, tropine and tetramethyl putrerscine. These alkaloids and valtropine occur in the mature root-bark.

Distribution

Duboisia grows wild on the borders and open woodlands along the east coast of Queensland and Northern and central New South Wales. It is indigenous also to New Caledonia. It is commercially cultivated in Australia, mainly in the Queensland area, from where it is exported to European. contries for further processing. It is also grown in Japan and India. Duboisia was first introduced in India at the Forest Research Institute, Dehra Dun and is being grown as a commercial crop in Karnataka, Andhra Pradesh, Madhya Pradesh and Maharastra.

Description of the Plant

D. myoporoides (2n = 60) is a shrub or small tree, occasionally attaining a height of 40 ft. When cut down or burnt, the plant produces

numerous suckers from the roots which form dense clumps. The bark is grey, corky, fissured, and the wood is light in weight. The leaves are alternate, 6–10 cm in length, petiolate, ovate-oblong to broad lanceolate, blunt-tipped and tapering at the base. The inflorescence axillary with terminal cymes, dichasial to begin with and showing a tendency to scorpioid cymes. The flowers are small, bell-shaped, white or pale-lavender or lavender-streaked, pedicillate and actinomorphic. The calyx has 5 persistent sepals. The corolla has five petals, is tubular and shows green streaks at the throat of the corolla. The androecium has 5 epipetalous stamens, the gynoecium is 2-celled with many ovules and axile placentation. The fruit is a berry, 5 mm wide, black and juicy when ripe, containing 5–6 seeds. The seed are small and kidney-shaped, one plant contains about 500 seeds. D. leichhardtii (2n = 60) is also a small tree or shrub reaching a height of 17ft. The leaves are generally shorter and more slender than those of D. myoporoides, 5–10 cm long, oblong-lanceolate or ovate, blunt-tipped, tapering at the base and entire.

The flowers are borne on terminal broad panicles, fairly leafy at the base, bell-shaped with flaring, narrow, long, pointed lobes much more conspicuous than the lobes in D. myoporoides' the calyx is small, with short, broad teeth. The fruit is oval in shape, 6 mm long with 6–12 seeds which are kidney-shaped and have a pebbly surface.

Varieties

There are no varieties available in this crop. However, a hybrid of D. leichhardtii and D. myoporoides has been developed. Higher alkaloid Yields are observed in crosses, where the female parent is D. myoporoides.

Soil

It can be grown on a wide range of soils, but it grows best in well-drained, medium to light-loamy soils, with a neutral pH.

Climate

It is a subtropical plant. Hence, it requires a moderate climate where the winter and summer temperatures are not severe. Areas with an average annual rainfall of 100 cm, well distributed throughout the year, are Ideal. Heavy rainfall is harmful to the crop.

Preparation of Land

Land is ploughed to a depth of 20–25 cm about 2–3 months before planting. The soil should be worked to a good tilth to eradicate weeds.

Cultivation

Propagation

Duboisia can be propagated from root-cuttings, terminal shootcuttings and seeds. But in commercial plantations, propagation by seeds or terminal-cuttings is practiced.

Raising the Nursery

Duboisia is propagated by seeds. The seeds are usually sown from February to April. The seeds are washed to remove all the pulp and dried in the sun before being used for propagation.

They are sown in sterilized soil in nursery beds under 60% shade, or seedlings can also be raised in polythene bags containing a mixture of fine sand and FYM in equal proportions, @ 2 seeds/bag. Pre-treatment of seeds with gibberellic acid reportedly gives better germination. The seeds are soaked in 250 ppm gibberellic acid at about 40°C for about 24 hours. The seeds are then thoroughly washed in fresh water, dried in the sun and can be used for sowing after about six weeks.

The seeds sown in nursery beds or polythene bags start germinating from about the 12th day and by the 18th day, about 10% of the seeds germinate. Nearly about 60% of the seeds germinate in about 40 days. Therefore, it is very essential to keep watering the nursery to ensure good germination. The seedlings raised in polythene bags are thinned, maintaining 1 plant/bag. The seedlings are ready for transplanting when they attain a height of about 25 cm and are 3 months old. About 250 seedlings are required for planting one hectare of land.

Transplanting

After the land is prepared, pits of 30×30×30 cm are made at 2m×2m spacing. The soil is mixed with the recommended dose of manure and fertilizers and made ready for transplanting. Healthy seedlings are planted in each pit, taking care to see that the seedlings are planted at the same depth as in the nursery bed or polythene bag, to avoid crown rot. Planting is usually done in the dry period. The field must be irrigated immediately after transplanting.

Manures and Fertilizers

At the time of land preparation, FYM @ 20 t/ha may be incorporated into the soil. Prior to transplanting, 20 kg N, 40 kg P_2O_5 and 49 kg K_2O/ha should be mixed well with the soil. Subsequently N @ 20 kg/ha is given to the crop at bimonthly intervals. Manure and fertilizers are supplied to the crop in the same dosage as above during the second and subsequent years of growth.

Irrigation

Plants should be watered frequently in the initial stages. The frequency is gradually reduced and when the crop is well established, irrigation once in about 15 days is sufficient during the dry period.

Interculture

Clean cultivation is a prerequisite; weeds are usually controlled mechanically. Grass-cover crops are sown after the leaf harvest.

Pests and Diseases

Pests

The major pests are the flea beetle, leaf-eating ladybird beetle, brown olive scale, trunk borer and green caterpillar. The crop can be sprayed with 0.2% Malathion whenever the insects are observed.

Diseases

No serious diseases have been reported in the crop. However, a viral infection causing leaf yellowing has been noticed in the plains of North India.

Harvesting and Processing

The plants are harvested after about 6 months of transplanting. The main as well as the side branches are cut at a height of about 1m, retaining about 20% of the leaves on the plant to encourage profuse branching. Two harvests are possible during the first year, and three harvests a year can be had from the second year onwards. The plants are likely to yield well for at least five years, after which, replanting should be done.

The leaves are separated and spread out for drying in the shade, which takes 2–4 weeks in warm weather, and are packed in gunny bags after drying.

Yield

The average yield of fresh leaves per harvest is about 2 kg/ plant which, on shade drying, reduces to about 0.5 kg. The fresh herbage yield is 10 t/ha in the first year and 15 t/ha from the second year onwards. This, on shade-drying, gives about 2 tonnes of air-dried leaves in the first year and 3 tonnes from the second year onwards.

Glory Lily

Plant Profile

Family	:	Wurmbaeoidae of Liliaceae
English name	:	Glory Lily
Indian name	:	Langali, Visalya (Sanskrit), Karihari Languli (Hindi), Agnishike, Gowrihoo Akkatangiballi (Kannada); Kalappaikkilanku Nabhikkodi (Tamil)

Species and varieties	:	Gloriosa superba Linn., G. lutea, G. plantii G. carsonii, G. richmondensis, G. virescens G. latifolia, G. magnifica, G. rothschildiana G. abyssinica, G. longifolia, G. simplex India, Africa, Asia, Indochina
Distribution	:	Madagascar, Sri Lanka, USA
Uses	:	Drugs

Gloriosa superba Linn., belonging to the family Liliaceae, subfamily Wurmbacoidae, is commonly known as glory lily, gloriosa lily or the lily flower. It is a fascinating genus of great scientific interest. In Kannada, it is called Agnishike, Akkathangiballi, Karadikannina gadde or Nangulika, while in Hindi, it is called Karihari.

The plant has been used in the Indian system of medicine since time immemorial. Its tubers are reported to have been used as a tonic, antiperiodic, anti helminthic and also against snakebites and scorpion stings. The drug is a gasterointestinal irritant and may cause vomiting and purging. It is sometimes used for promoting labour pains and, conversely, also as an abortifacient. It is considered useful in colic, chronic ulcers, piles and gonorrhoea. It is used in local applications against parasitic skin diseases and as a cataplasm in urological pains. The leaves, when applied in the form of a paste to the forehead and neck, are reported to cure asthma in children. The leaf-juice is used against head lice.

The medicinal properties of the drug are due to the presence of alkaloids, chiefly colchicine ($C_{22}H_{25}O_6N$;mp 151–52°) and gloriosine ($C_{22}H_{25}O_6N$;mp 248-50°). Colchicine is used in the treatment of gout, a common disorder in the temperate parts of the world. Gout is caused by the deposition of micro-crystals of uric acid in the joints, which is attributed to a defective regulatory mechanism for endogenous passive synthesis. It is said that colchicine interrupts the cycle of new crystal deposition which seems to be essential for the continuance of acute gout. In addition, these alkaloids are also used as polyploidizing agents in polyploid breeding in crop research. The colchicine content in Gloriosa superba Linn. varies from 0.15 to 0.3% in the tubers, and in the seeds it ranges from 0.7 to 0.9%. Until recently, only the tubers are exploited but, after discovering the fact that the seeds also have a high quantity of alkaloids, the crop is grown mainly for its seeds which are in great demand within the country and in the international market.

Origin and Distribution

Gloriosa is a native of tropical Asia and Africa. The genus derives its name from the Latin word gloriosus, referring to the flowers. It is found growing throughout tropical India, from the North-west Himalayas to Assam and the Deccan peninsula, extending up to an elevation of 2120 m. In Karnataka, it is commonly found growing all along the Western Ghats. It is also found growing in Madagascar. Sri Lanka, Indochina and on the adjacent islands. The area under this crop in India is around 2000 ha.

Related Species

The genus Gloriosa is comprised of about 10 to 15 known species. The important species found in India are G. superba and G. rothschildiana. The other species found growing in other parts of the world are G. carsonii, G. plantii (USA, Africa) G. simplex, G. abyssinica (Africa), G. richmondensis, G. virescens, G. lutea, G. magnifica, G. longifolia and G. latifolia. Gloriosa is monobasic with a genetic base x = 11. Out of the 10 elemental species, G. superba, G. lutea and G. plantii are diploids (2n = 22); G. carsonii, G. virescens and G. richmondensis are tetradploids (2n = 44) and G. rothschildiana, G. latifolia and G. magnifica are octoploids (2n = 88). In general, octoploid species are comparatively short-statured and constitute a medium group of plants.

Description of the Plant

It is a herbaceous, climbing perennial, growing between 3.5 to 6 m in length, but usually trained at 1.5 m above ground-level. The vines are tall, weak-stemmed with tuberous roots that support themselves by means of cirrhosed tips. The leaves are ovate, lanceolate, acuminate, the tips spirally twisted to serve as tendrils. The flowers are large, solitary or may form a lax-corymbose inflorescence, twisted and crisped with six recurved or reflexed petals, blossoming yellow but changing to yellow-red and deep scarlet. In the bud stage, the petals hang down over the ovary and, on maturity, they assume an erect position, leaving the ovary with its stigma exposed at right angles. There are five, long stamens, each with a long anther that displays profuse orange-yellow pollen. The ovary is 3-celled and it forms an ellipsoidal capsule. There are numerous seeds in a capsule and the seeds are warty and dorsally compressed.

Crop Improvement

In the year 1979, a cross-breeding programme was attempted involving G. superba, G. lutea and G. plantii (diploids), G. carsonii, G. virescens and G. richmondensis (tetraploids) in several combinations. The only successful hybrids were G. virescens × G. carsonii and its reciprocal: G. virescens × G. richmondensis and G. carsonii × G. richmondensis. Morphological analysis and cytogenetical behaviours of all the hybrids has been worked out. Based on these studies, it was reported that the tetraploid Glorioso species had higher crossability as compared to the diploid species. Furtner, G. superba has always failed to produce hybrids, although it was involved in numerous hybridization combinations.

Soil

It prefers sandy loam soils on the acidic side, with good drainage, for its successful growth. In Southern India, it is found growing luxuriantly in red and black loamy soils which have a good amount of humus with a medium water-holding capacity. A soil pH range between 6–7 has been found suitable for raising this crop.

Climate

Glory lily is a tropical plant and comes up well in warm, humid regions. Under natural conditions, it is found growing up to an elevation of 600 m from sea-level. An annual rainfall of about 373 cm, well-distributed throughout the year, is ideally suited for this crop. It does not withstand continuous moisture stress and requires frequent ii ·igations up to the flowering in dry periods. The temperatures favourable for its growth and flowering are 15–20°C during the day and 10–15°C at night. The relative humidity should be high. However, continuous cloudy weather is congenial for the pathogen Curuularia, which is a very serious threat, resulting in 75 to 80% mortality of the vines. Suitable prophylactic measures have to be adopted, in case the crop has to be grown under such conditions.

Cultivation

Propagation

It is commercially propagated from its underground, V-shaped rhizomes or sexually propagated by seeds. The plants raised from seeds take nearly three to four years to flower. Hence, except for experimental purposes, seed propagation is not favoured by the growers.

Gloriosa produces a biforked tuber during the growing season and each of these forks has only one growing bud. Tubers should be handled carefully, as they are brittle and liable to break easily. If the growing bud is subjected to any kind of damage, the tuber will fail to sprout. Since the vigour of the vine and its flowering and fruiting ability depends on the size of the tubers, it should not weigh less than 50–60 g. The plants raised from smaller tubers do not produce flowers during their first year. Large tubers can be divided into two by breaking them in the middle. The dormant tubers start sprouting from the month of May. Under Bangalore conditions, planting during the months of July and August has been found to favour good growth and yield. About 2.5 to 3.0 t/ha of tubers are required for planting. In order to avoid rotting of the tubers before sprouting, only healthy tubers should be selected for planting. In addition, the selected tubers or tuber pieces should be treated with suitable fungicides, preferably Emisan-6 (methoxy ethyl mercury chloride) @ 0.08%.

Field Preparation and Planting

The field should be ploughed and harrowed several times until it is brought to a fine tilth. All the grass-stubble and roots should be removed. The field must be levelled properly and drainage arrangements made to avoid water-logging during the rains. The field is then divided into subplots of convenient sizes. About 15–20 t/ha of FYM or compost should be mixed well into the soil. One foot deep furrows are opened at a spacing of 45–60 cm. The treated tubers are planted at a depth of 6-8 cm, keeping a plant-to-plant distance of 30 to 45 cm, depending upon the type of soil. Closer spacing has been reported to favour cross-pollination, thereby improving the fruit set.

Manures and Fertilizers

Though gloriosa makes satisfactory progress with little manuring and fertilization, the addition of well-decayed manure, bonemeal and fertilizers to the soil ensures a vigorous plant, stronger tubers and better flowering. Experiments carried out on this crop have shown that under Bangalore conditions, a fertilizer dose of 120 kg N, 50 kg P_2O_5 and 75 kg K_2O/ha is required for a good crop. Of the nutrients, the whole P_2O_5 and K_2O and one-third of N is applied as a basal dose and the remaining two-third of N should be given in the first six to eight weeks after planting.

Irrigation

Frequent irrigation is required during the sprouting time to keep the surface soft, so that there is no hard-pan formation, in order to facilitate easy sprouting and emergence of the growing tip outside the soil. Irrigation should be withheld until after the flowering is over, to prevent rotting of the tubers. Excess watering is harmful to the plants and causes yellow or brown-coloured patches on the leaves, which fall off prematurely.

Crop Monitoring

The provision of some kind of support is necessary for successfully growing gloriosa. Since the stem is very tender, when the plants are about 30–40 cm tall, they should be staked or tied to wires or allowed to climb on some sort of frame.

Weeding

In the initial stages, the gloriosa plantation requires frequent weeding to control the weeds which will otherwise compete with plants for moisture and nutrients and will restrict the growth of the plant. While weeding, utmost care should be taken to avoid any damage to the growing tip, as once damaged, it does not sprout again during the season. Chemical weed control is possible only when there is wide spacing between the rows and the plants themselves.

Pests and Diseases

Comparatively speaking, gloriosa has few pests and diseases. However, some pests and diseases pose a serious threat when they get favourable conditions, causing severe damage. Hence, great care has to be taken to control them.

Pests

1. Lily caterpillar (Polytela gloriosae)

The lily caterpillar, marked with bands of red and black colour on its body, attacks almost all parts of the plant. They lay a mass of eggs on the undersurface of a tender leaf or in the crown of the growing shoot. Within a very short time, the newly-hatched caterpillars with their voracious appetites damage the leaves and growing points. This pest can be effectively controlled by spraying Metacid at a concentration of 0.2% at fortnightly intervals.

2. Green caterpillar (Plusia chaleites)

This caterpillar is small and green in colour and attacks the leaves and flowers. It completely eats away the flowers causing severe damage to the crop. The control measures recommended for the lily caterpillar will control this pest also. In addition to the above two insects, the following twelve new insect pests have been reported from the University of Agricultural Sciences, Bangalore. The new pests are: Diabolocatantops pinguis, Oxya juscouittata, Monolepta signata, Curculio spp., Myllocerus udecimpunctatus, M. uiridanus, Ptochus spp., Amsacta lactinea, Utetheisa pulchella, Archips micaceana, Lygaeus hospes and Graptostethus seruus.

Diseases

1. Leaf Blight

This disease is caused by the fungus Curuularia lunata. The spores of the fungus are airborne and attack the foliage. The affected leaves first turn light-yellow and, later, small black spots are seen which ultimately spread to the entire leaf and neighbouring plants. The incidence is higher during cloudy weather coupled with high humidity. The disease can be controlled by spraying 0.3% Dithane M–45.

2. Tuber rot or Basal stem rotting and wilting

The soil-borne Sclerotium spp. affects the underground tubers and causes the death of the plant. In the initial stages, infected tubers start becoming soft and the foliage exhibits a yellow appearance and in advanced stages, the whole tuber gets infected giving the appearance of a discoloured mass and the plant dies off. Drenching the soil with Bavistin @ 0.2% or Cuprassol near the root-zone of the plant has been observed to control the disease.

Harvesting and Processing

Gloriosa is a crop of 170-180 days duration. When planted in June, it starts bearing flowers after 55 days and continues to flower and fruit till October. The fruit requires about 105-110 days from the set to reach maturity. The right stage of harvest is when this capsule starts turning light-green from dark-green and the skin of the fruit shows a shrunken appearance and becomes light in weight. At this stage, when pressed, the pod gives a crinkling sound. After picking, the capsules should be kept in the shade for 7 to 10 days to facilitate the capsules to open up, displaying deep orange-yellow coloured seeds. The seeds and pericarp are separated manually and dried for a week in the shade, by spreading them uniformly over any clean, dry floor or any platform specially erected for the purpose. At the later stages, the seeds are moved to the sunlight for a week till they dry completely. The dried seeds are then packed in moisture-proof containers and stored until exported or extracted for the alkaloids.

Yield

The yield of seeds differs greatly, depending upon the vigour and age of the plant which, in turn, depend on the size of the tuber. The yield in the initial year will be low, but it gradually increases in the subsequent years. After three years, from a well-managed field, we may get about 200–250 kg/ha of dried seeds. The yield of pericarp (husk) is about 75% of the seeds.

Chemical Analysis of Seeds

Colchicine has been extracted from the corms of Colchicum automnale, growing wild in some parts of Europe. Recently, the supplies of colchicine from the conventional sources has not been sufficient to cope with its increasing demand. Among the other Indian plants, the seeds of Iphigenia stellata contain 0.90%, the corms of Colchicum leuteum contain 0.25% and the seeds of Gloriosa superba were found to contain 0.60% of colchicine. The easy availability from both wild and cultivated sources make the seeds of Gloriosa superba a potential source of colchicine in India.

In order to extract colchicine, 20 g of powdered seed material is exhaustively extracted in a Soxhlet's apparatus with 150 ml of methanol. The extract is concentrated to about 10 ml, when a white deposit is formed. On TLC, it is found that colchicine is present in the mother liquor and not in the white deposit (0.6 g). The mother liquor is dried in a vacuum to get 3.7 g of extract. One gram of the extract is taken as a preparative for thin-layer chromatography. The extract is spotted on precoated silica gel plates, every time using pure colchicine as a standard on the left-hand side. After detection in the iodine chamber, the band corresponding to colchicine was scraped, extracted with methanol and dried (0.03 g). The colchicine obtained, corresponds to 0.55 % of the dry seed material.

Dvovackova's Metod

About 20g of powdered seed is extracted with methanol in a Soxhlet's apparatus. The extract is concentrated and diluted with water and is acidified to pH4, with citric acid. It is extracted with ether and the ether extract is discarded. Further extraction with chloroform gives the neutral phenolic portion which, on passing through a column of silica gel and drying in a vacuum, gives almost pure colchicine.

Guggal

Plant Profile

Family	:	Burseraceae
English name	:	Indian Bedellium tree
Indian name	:	Gugguluh, Mahisaksah (Sanskrit) Gugal, Guggul (Hindi) Gukkulu, Mahisaksi (Tamil)
Species and varieties	:	Commiphora agollochoa Engl C. stocksiana Engl C. mukul Hook ex stock C. berryi (Aen). Engl Anand, Marusudha
Distribution	:	India, Africa, Asia, Australia, Bangladesh Pacific Islands, Pakistan

Uses : Drugs, Perfumery

Guggal or the Indian Bedellium tree is Commiphora mukul (Hook ex Stocks) Engl. Syn. Balsomodendron mukul, a small tree belonging to the family Burseraceae. This plant is a source of Indian Bedellium, a oleo-gum-resin obtained by incision of the bark. The resin occurs in vascular or stalactitic pieces, paleyellow, brown or dull-green in colour with a bitter aromatic taste and balsamic odour. The oleo-gum-resins are mixtures of resin (about 61 %), gum (about 29.3%), volatile oil (about 1.45%) and occasionally combined with other substances, mainly myrcene, dimyrecene and polymyrecene. It has ketonic and non-ketonic fractions. The ketonic mixture constitutes about 12% of ethyl acetate soluble neutral part which consists of 8 steroids. Two of these steroids, Z-and e-guggalsterone, constitute 2% of the gum resin and are responsible for the lipid-lowering activity of guggal. Non-ketonic fractions exert a synergistic action on the biological activity of the ketonic fraction. The ethyl acetate extract of guggal has been named 'guggulipid'.

The resin is largely used as incense and as a fixative in perfumery and in medicine. In indigenous medicine, it is used as an astringent, antiseptic, stomachic, carminative and digestant. The oleo-resin causes an increase of leucocytes in the blood and stimulates phagocytosis. It acts as a diaphoretic, expectorant and diuretic and is said to be a uterine stimulant and an emmenagogue as well. It is highly effective in the treatment of obesity, arthritis, indolent ulcers, weak and spongy gums, pyorrhoea, alveolaris, chronic tonsilitis and pharangytis, ulcerated throat, and chronic dyspepsia. Inhalation of the fumes of burnt guggul is recommended in hay fever, acute and chronic catarrh, chronic laryngitis, chronic bronchitis and phthisis. It is an ingredient of ointments for ulcers.

Origin and Distribution

The genus Commiphora, which has 165 species, has its origin in Africa and Asia and is widely distributed in the tropical regions of Africa, Madagascar, Asia, Australia, Pacific Islands, India, Bangladesh and Pakistan. Four species occur in India. These are: C. agollochoa Engl; C. stocksiana Engl; C. mukul Hook ex stock and C. berryi (Aen). Engl. The genuine guggal gum is derived from C. mukul. In

India, this species is distributed in the states of Rajasthan, Tamil Nadu, Assam, Gujarat, Maharastra and Karnataka.

Description of the Plant

A small tree or shrub up to 3–4 m high, the branches are crooked, knotty, aromatic and end in sharp spines. The bark is papery and peels in strips from the older parts of the stem. The leaves are sessile, alternate or fascicled, 1–3 foliate, the leaflets are glabrous. The terminal is sessile or subsessile, obvate, serrate, 1–5 cm long, 0.5 to 2.5 cm broad, lateral, when present, sessile, serrate or entire. The plants are dimorphic, one having bisexual and male flowers and the other having female flowers with staminodes. The bisexual and male flowers are sessile, 3-5 mm long, usually red and rarely pinkish-white. The flowers appear in groups of 2 or 3. The calyx is fused with the disc basally, it is tubular or urceolate, 1-2 mm long. The petals are reflexed, acute, 3–5 mm long and 1 mm broad. There are 8 stamens, very rarely 10; 3–5 mm long. The disc is conspicuously toothed. The ovary is 2-loculed with a sessile 2-lobed stigma. The fruit is ovoid, up to 1 cm-long drupe, red when ripe and the mesocarp is yellow. When it is ripe, it splits into two.

Varieties

About 26 cultures designated by the name of the places from where they are collected are being evaluated at Anand, in Gujarat, from 1981 and a variety named 'Marusudha', which is the highest yielder, has been released for cultivation.

Soil

Guggal is not grown on a commercial plantation scale anywhere. But the plants are found growing naturally in arid tracts of Western India over sandy to silt-loam soils, which are poor in organic matter but rich in several other minerals. Faster growth of the plant is observed in soils which have moisture-retaining capacity. An average soil is suitable for its cultivation.

Climate

The plant has a wide adaptability and is found growing in arid regions under varying conditions. Its small, leathery leaves and thick bark, covered with a white waxy coat over the stem, helps it to withstand desert conditions. It prefers a warm, dry climate for a good yield of oleo-gum-resin.

Land Preparation

The land is prepared well in advance of the rainy season by 2–3 ploughings and laid out into plots of convenient sizes. Pits of the size 0.5×0.5×0.5 m^3, are dug at a spacing of 3 to 4 m, in rows. The pits are filled with FYM and topsoil and mixed with Aldrin (5%), to prevent any damage to the plants due to termite attack.

Cultivation

Propagation

Guggal can be propagated by seeds and vegetatively through stem- cuttings. Air-layering has also been successful in this plant. In literature, the protocol for meristem culture is also available.

(a) Propagation by Seeds

This is not a common method of propagating this plant, because the seed germination is slow and is also very poor (5%) due to the presence of the hard seed-coat. To achieve a good rate of germination, the seeds are mechanically scarified with sandpaper and are kept under running water for 24 hours. The seedlings may be raised in polythene bags during the kharif season and, after hardening, may be planted in the main field.

(b) Propagation by Stem-cuttings

About 15–20 cm long and 10 mm thick, semi-hardwood cuttings are taken and the cut end is treated with IBA or NAA growth-regulator solutions and planted in well-prepared and manured nursery beds during the months of June and July. The beds are irrigated lightly after planting and regularly thereafter. The cuttings sprout in 10-15 days, they grow well and are ready for planting in the main field after 10 to 12 months, during the next rainy season. The percentage of rooting in stem-cuttings is around 80 to 94%.

Planting

The rooted cuttings are planted in the pits during the rainy season. As the plants grow, they are trained properly by cutting the side branches.

Fertilizer Application

This crop has not shown good response to fertilizers, except to low-level or nitrogen fertilization. Hence, urea or ammonium sulphate at 25/50 g/bush is given twice a year before irrigation.

Irrigation

Providing light irrigation during the summer season supports good growth of the plants.

Intercultivation

Interculture is confined to one weeding and hoeing in the early stages of growth. But the soil has to be stirred around the bushes twice in a year. This practice of soil-stirring is found to increase the growth of plants.

Pests and Diseases

Guggal plants are attacked by a leaf-eating caterpillar Euproctis Janata Walker; white fly Bemisia tabaci, which sucks the sap of leaves as a result of which the leaves turn yellow and finally drop off. The rooted cuttings also suffer from termite attacks in the early stages of growth.

The diseases noticed on this crop are leaf-spot caused by Cercospora spp., and bacterial leaf-blight.

Gum-tapping and Yield

The plants attain normal height and girth after 8 to 10 years of growth, when they are ready for tapping. For tapping the gum, which is present in the balsam canals in the phloem, bark-deep, i.e., a shallow, incision is made on the bark. If the incision is too deep, the plant either dies or yields little resin in the following years. While making the incision, a small quantity of guggal gum mixed with water may be applied to the incised place using a prick-chisel. The sharp end of the chisel is dipped in the guggal solution and an incision is made on the bark carefully. Usually, the incision is made after November but before April. The resin is collected at an interval of 10-15 days. Weather conditions influence the success of obtaining gum. The yield of gum-resin from each plant is about 700–900 g.

Separation of Resin from Gum

The resin can be separated from the gum either by hot expression or solvent extraction at 120–130°C. The yield by the former process is about 10% less than that obtained by solvent extraction (61%). The purified resin is transparent in thin films, but, translucent or even opaque in bulk. It is completely soluble in most of the organic solvents and in castor oil, drying oils and turpentine and mixes well with vegetable waxes, stearic acid and resin.

Pruning of Plants

In experiments conducted by CIMAP, Lucknow, on guggal plants obtained from the arid regions of Gujarat and Rajasthan, it has been demonstrated that guggal stems pruned in May contain the maximum amount of guggal sterons (0.06%). This method saves the plant from drying and ultimate death, which was a result of the extraction procedure adopted earlier.

Henbane

Plant Profile

Family	:	Solanaceae
English name	:	Henbane, Black henbane
Indian name	:	Parasikayavani (Sanskrit) Khurasani ajavayan, Khurassani jamani (Hindi) Kurassaniyomam (Tamil)
Species and varieties	:	Hyoscyamus niger L.

		H. muticus L.
		IC-66
		Aela
Distribution	:	India, Amenca, Central Asia, Europe Finland, Greece, Portugal From Western Punjab through Afghanistan to Egypt
Uses	:	Drugs

Henbane, comprising of the dried leaves and flowering tops of Hyoscyamus niger L., commonly called black henbane or common henbane and H. muticus L., is an official drug of repute. The herb and seed yield tropane alkaloids (0.03–0.06%), 90% of which are hyoscyamine and hyoscine and the rest are scoplamine, atropine, hyoscipirin, choline, fatty oil, mucilage, albumen and KNO_3. They are used as an anodyne, sedative, tranquillizer, antiseptic, antispasmodic, mydriatic, in asthma and whooping cough. Hyoscine is particularly useful in protection against shocks caused by accidents and loud noises.

Origin and Distribution

Hyoscyamus niger is a native of Europe, Central Asia, India and tropical America and is distributed throughout Europe from Portugal and Greece in the South to Norway and Finland in the North. In India, it is mainly found growing in temperate to subtropical regions and is grown largely in Kashmir and Uttar Pradesh. Its successful cultivation at Indore, recently, has indicated that it can also be grown in warmer regions as a winter crop. The plant has been found to grow well under Bangalore conditions also. H. muticus is, however, essentially a desert plant attaining a height of 90 cm (and 30 cm spread) and is found growing from Western Punjab through Afghanistan to Egypt.

Description of the Plant

H. niger (2n = 34), belonging to the family Solanaceae, is an erect, viscidly hairy, biennial plant growing up to 160 cm in height, and is grown as a short duration (90–100 days) winter crop in the Indian plains. The leaves are simple, large, ovate or oblong, coarsely dentate and pinnately lobed. The flowers are yellowish, pale, sessile or subsessile, borne on terminal cymes. The fruit is a berry containing numerous minute seeds. The seeds are small, about 1.5 mm long, 2000/g, brown to black, marked with fine, conspicuous reticulations.

H. muticus Linn (2n = 28) is very similar to H. niger, but shorter in height (90 cm). The leaves are small, having a smooth dentate margin with a thick and sticky texture. The seeds are somewhat heavier, covered with a dark brown seed-coat.

Varieties

IC-66 is a short duration (100 days) variety of H. muticus, developed and released by the National Bureau of Plant Genetic Resources, New Delhi. It is reported to yield 25 q/ha with the total tropane alkaloids varying from 0.05 to 0.10%. Another variety which is a tetraploid and a profusely-seeding polyploid line has been released by the CIMAp, Lucknow. This variety is reported to have a growing period of 200 days and produces 645 q/ha fresh herb or 43.2 q/ha dry herb, in comparison to 511 q/ha and 37.3q/ha in the diploid parent. The alkaloid content is also high, suggesting that it will be more profitable than the existing varieties. The CIMAp, Lucknow, have released one variety of H. niger through mutation breeding. Aela is the mutant culture of H. niger, selected from irradiated progenies, which have yellow flowers with a slight purple tinge at the base of the petals. It grows vigorously giving a yield of 73 q/ha, which is 109% more than the parent (50 q/ha). It has a high total of alkaloids-0.545% against 0.167% in the parent plant.

Soil

Henbane, being a foliage crop, requires a level, well-drained fertile, sandy loamy soil, rich in organic matter, for its optimum growth. A soil pH range of 7–8.5 is preferred by the crop. Deep black and water-logged soils are not suitable for its cultivation.

Climate

Henbane is a cool season crop and it does not tolerate very cold temperature, as the growth and development of these plants is appreciably slow below 20°C and becomes almost static if the temperature remains below 10°C. The plants perish under frosty conditions. Low temperature has also been reported to affect the seed germination. Seeds of H. niger and H. muticus germinate best in 8–10 days if a temperature between 20–30°C is maintained with adequate moisture. In temperate regions, it is raised as a summer crop whereas in subtropical regions, it is cultivated as a winter crop. In any case, the plantation must get bright sunlight which is important for the proper growth and development of this crop. Similarly, regions with high temperature and humidity must be avoided as both are detrimental to plant growth and also to their content of alkaloids.

Cultivation

Propagation

The crop can be raised either by sowing the seeds directly in the main field or by raising the seedlings in the nursery and transplanting them.

Successful regeneration of plants from hairy root cultures has been achieved in H. muticus through hormonal induction,at the CIMAP, Lucknow.

Direct Sowing

The best time for sowing the seeds directly in the field is early October in subtropical regions, while June-July is most appropriate in temperate regions. As henbane seeds are very small, it is important to ensure that the soil is prepared to a fine tilth for sowing. Seeds must be sown not deeper than 2 to 2.5 cm below the soil for good germination. Soaking the seeds overnight and incubating them for 7-8 days at 25°C for sprouting before sowing, gives a better germination rate. In order to ensure the uniform distribution of seeds all over the field, the seeds should be mixed with dry, powdered earth or sand in a ratio of 1: 10 before sowing. About 2–3 kg of seed would suffice to sow an area of one hectare.

The optimum sowing time of H. niger at Indore is reported to be the second fortnight of November with 30 × 30 cm spacing. In Delhi, the first fortnight of November, with 30 × 15 cm was found suitable.

Raising of Seedlings

The best time for raising the nursery is September in the plains and April-May in temperate regions. The seeds are mixed with sand and sown by broadcasting in rows of 8 cm, in raised nursery beds. The seeds are then gently covered with a fine mixture of soil and FYM or sand, mulched and immediately watered. About 500 g seeds are required to raise enough seedlings to cover an area of one hectare. Germination takes place in about 8–10 days. At this time, the mulch is removed but watering and weeding are continued as usual. If the temperature is low, the seeds take a longer time to germinate and the growth of the seedlings is also very slow. The seedlings take about two months to reach a suitable stage for transplanting. A foliar spray of 1% urea, when seedlings are about 20 days old, has been found to accelerate their growth.

Planting

The seedlings are transplanted at a spacing of 45 cm in rows and 15–30 cm from plant-to-plant in a well-prepared field and followed up with light irrigation. Transplanted seedlings take about a week to establish. Since there is a heavy mortality in the transplanted seedlings, it is recommended that this crop should be grown on a commercial scale only by direct sowing.

Manures and Fertilizers

At the time of land preparation, well-decomposed FYM at the rate of 15–20 t/ha is added. In addition, a dose of 40–80 kg N, 30–40 kg P and 20–30 kg K/ha, has been recommended to obtain the maximum dry-matter yield. The entire quantities of P and K along with 20 kg N is given as a basal dose. The remaining quantity of nitrogen is given as a top-dressing in two equal split doses during the growing season.

Irrigation

The transplanted crop is given the first irrigation immediately after transplanting. Subsequent irrigations are given at 8–10 days' intervals, until it is about a month old and, subsequently, the crop is irrigated at intervals of about 15–20 days. In all, 4–5 irrigations are required. But in the directly sown crop, to secure the proper seed germination, it is necessary that one pre-sowing irrigation is given before ploughing. If the seed germination is sparse, a week after sowing another irrigation is essential. After this, 3–4 more irrigations at an interval of 15–20 days are required for the standing crop.

Interculture

The crop raised by direct sowing should be thinned after a fortnight of germination and a spacing of 30 cm within the rows should be maintained. Usually, two weeding-cum-hoeing operations are necessary during the entire cropping season.

Pests and Diseases

No serious attack of insect pests and diseases are reported on both the species of henbane. However, the larva of the cotton bug attacks the leaves and capsule of this crop during July-August, causing heavy damage. Aphids also infest the crop. These two insects may be controlled by 2–3 sprayings (at 10–15 days' interval) of Endosulphon (2 ml/l of water) or Methyl Parathion or Metasystbx.

A green mosaic caused by a virus has been reported to affect the crop in India. The nematodes Meloidogyne incognita and M. jauanica cause root-knot disease.

Harvesting Processing and Storage

The leaves and flower-tops of H. niger are collected at the flowering stage of the plant, while the entire aerial part of H. muticus may be harvested at a similar stage. In H. niger, the older leaves of the lower regions of the plant touching the ground should be picked first and dried separately. The herb is sun-dried for 2-3 days followed by 6–7 days in the shade, with constant raking over with sticks. Approximately, every 100 g of fresh leaves yield 14 g of dry leaves. After drying, the material is stored in gunny bags and kept in a cool, dry place. The produce should not be stored for more than 2 months.

Yield

The yield of the herb varies greatly with the spacing. By adopting a spacing of 45×30 cm, an average yield of 15q/ha of dry herb may be obtained.

Estimation of Hyoscyamine in Leaves

For this purposes, 40 g of leaf-powder is weighed and transferred to a flask containing 200 ml of a mixture of 4 volumes of solvent and 1 volume of ethanol. It is shaken well and allowed to stand for ten minutes. Then 6 ml of dilute ammonia solution is added and it is shaken frequently for one hour. The mixture is then transferred to a percolator plugged with cotton-wool and, when the liquid ceases to flow, packed firmly. Continue the percolation, first with 100 ml of the ether-alcohol mixture and then with solvent ether, until complete extraction of the alkaloids is effected.

The total time of percolation should not exceed three hours. The volume of the percolate is reduced to about 20 ml by distilling over a water-bath. The distillate is transferred to a separator, then the flask is washed with chloroform three times taking 10 ml each time and the washings are added to the separator followed by 30 ml of 0.5 N sulphuric acid. The separator should be shaken well, allowed to separate, and the lower layer allowed to run off.

The extraction is continued with successive quantities each of 10 ml of a mixture of 3 volumes of 0.1 N hydrochloric acid and 1 volume of ethanol, until complete extraction of the alkaloids is effected. The mixed acid solution is washed with about 10 ml of chloroform and the latter run-off into a second separator containing 20 ml of 0.1 N hydrochloric acid.

It is shaken, allowed to separate and the chloroform is discarded. Repeat the extraction in the first separator with chloroform, twice, taking 5 ml each time, and transfer the liquid to the second separator each time and wash with the same aqueous acid as before. Transfer the acid liquid from the second separator to the first separator. The liquid is made distinctly alkaline with a dilute ammonia solution (25%) and shaken with successive quantities of chloroform, until complete extraction of the alkaloids is effected. Wash the combined chloroform solution with about 3 ml of water. Remove most of the chloroform and transfer the remainder of the chloroform solution to a shallow, open dish.

Complete the removal of the chloroform, add 2 ml of ethyl alcohol, evaporate to dryness at 105°C, and weigh at intervals of one hour until two successive weighings do not differ by more than 1 mg. Dissolve the residue in 20 ml of 0.02 N sulphuric acid, and titrate with 0.02 N sodium hydroxide, using a solution of methyl red as indicator. Each ml of 0.02 N sulphuric acid is equivalent to 0.005787g of alkaloids, calculated as hyoscyamine.

Holy Thistle

PLANT PROFILE

Family	:	Asteraceae
English name	:	Holy Thistle
Indian name	:	Pavitramullina Gida (Kannada)
Species and varieties	:	Silybum marianum Gaertn
Distribution	:	India, Africa, Asia, North and South America Europe
Uses	:	Drugs

It is an erect annual or biennial, growing up to 120–130 cm tall. The stem is pale-green, simple or slightly branching, the leaves are large, mottled with white, pinnatified into ovate, triangular, sinuate-toothed, spiny lobes. The heads are globular, 6–10 cm broad and concave at the base. The outer scales of the involucre are oblong, broadening into an ovate, prickly, cellate margin appendage which tapers abruptly into a long spine, the inner scales are lanceolate and entire. The flowers are hermophrodite and fertile, purple or white, the tube is slender, the limb is dilated below and is five-fold.

Soil

It is a very hardy plant and is found growing on a variety of soils. However, sandy loam loam soils which have good drainage are ideal. The crop can be grown successfully even on marginal lands by supplementing the nutrients through fertilizer application.

Climate

The holy thistle prefers a moderate type of climate for its successful growth. Places with extreme climates are unsuitable for its cultivation. The plant can be grown up to an altitude of 2400 m from sea-level.

Cultivation

Sowing

The crop is raised by using the seeds obtained from the previous season's crop. It is sown during September-October in the plains and March-April in hilly areas. Before sowing the seeds, the field is prepared well, divided into convenient sized plots and throughly mixed with organic manure. Then the seeds are either sown by broadcasting them in the rows spaced at 60 cm and later thinned to maintain an approximate spacing of 60 cm between the plants within the row, or they can also be sown at 60×60 cm spacing using a seed-drill. If enough moisture is not available, the field should be irrigated immediately after sowing.

Manures and Fertilizers

The crop responds well to the application of manure and fertilizers. FYM at the rate of about 8–10 t/ha should be added at the time of land preparation. In addition, a dose of 120 kg N, 100 kg P_2O_5 and 75 kg K_2O/ha, is recommended for obtaining a good yield. Among the fertilizers, a half-dose of N and the whole dose of P and K are applied at the time of planting and the remaining half-dose of N is given as a top-dressing after the seedlings are one month old.

Irrigation

It is a sufficiently hardy crop and does not require very frequent irrigations, except during the early stages of growth, when the crop has to be irrigated at weekly intervals. Subsequently, watering once in a fortnight will suffice. In all, about 10–12 irrigations during the entire cropping season will enable the crop to grow luxuriantly.

Interculture

Weeds are a problem in this crop only during the early stages of growth. Once the crop picks up, it covers the field with its large foliage and the weeds are automatically supressed. One or two weedings when the plants are young will help to avoid competition from weeds.

Insects Pests and Diseases

Silybum is practically free from most pests and diseases. However, there is a report of the pest Tanyameus polliatus which can be controlled by a spray of 0.03% Methyl Parathion. Among the diseases, root-wilt caused by the Rhizoctonia species has been observed under Bangalore conditions, which can be controlled by drenching the soil in fungicides like Captan or Brassicol (0.1 %)

Harvesting and Yield

Silybum is a 4 months duration crop. The crop is prickly and hence it is to be harvested carefully by cutting the individual heads with the help of a sharp instrument. Since all the heads do not reach maturity at the same time, harvesting is done 2–3 times. There is also a problem of seed-shattering; therefore, the heads should be harvested when the seeds are just turning brown. If the crop is to be harvested mechanically, it should be harvested when the majority of the heads have reached maturity. The heads are spread in the threshing-yard for 3–4 days for drying. While drying, quite a few of them burst and release the seeds, and the remaining are threshed, separated and stored. If harvested at the right time, on an average, about 1 t/ha of seeds may be obtained.

Honey Plant

Plant Profile

Family	:	Apiaceae
English name	:	Honey Plant
Indian name	:	Madhugida (Kannada)
Species and varieties	:	Amm; majus L. A. visnaga Suttons, Monoica
Distribution	:	India, Abyssinia, west Africa, Egypt, Europe France, Germany, the Mediterranean Region, Switzer Land
Uses	:	Drugs

Honey plant (Ammi majus L.), belonging to the family, Apiaceae; Umbelliferae, is an important medicinal plant indigenous to Egypt. Its seeds constitute the principal commercial source of xanthotoxin. The other constituents of the seed are: bergapten isopimpinellin, isoimperatorin, oxypeucedanin, heraclenin, oxypeucedanin hydrate, sexalin, pabulenol, and three other compounds, and the synthetic psoralens: xanthotoxol, bergaptol, imperatorin, heracelin, heraolenol and sexalin. Around 16 psoralens have been obtained from A. majus; thus, this species is one of the richest known source of linear furocoumarins. Xanthotoxin is commonly used in the treatment of vitiligo or leucoderma and in formulations of suntan lotions. The xanthotoxin content in the fruit is about 0.4% and additional quantities of it can be prepared by converting the imperatorin present in the fruits by chemical means. The plant is reported to be biennial in its natural habitat, but behaves as an annual under cultivation in India.

Distribution and Trade

The plant is indigenous to Egypt and is widely distributed in Europe, the Mediterranean region, Abyssinia and West Africa. The plant is reported to have been first introduced in India along with A. visnaga by the Forest Research Institute, Dehradun, in 1955, by the courtesy of the UNESCO. It is being successfully grown in the temperate and subtropical regions of Himachal Pradesh, Uttar Pradesh, Gujarat, Tamil Nadu and Karnataka. Although the recent statistics on the area, production, import and export details of this crop are not available, but according to Singh (1963), France, Switzerland and the Federal Republic of Germany alone require about 400 t/yr of Ammi majus fruits and part of this requirement is being met by the exports from India. On the other hand, India has been importing xanthotoxin worth about two lakhs annually.

It is an annual herb attaining a height of 80 to 120 em. It has a long taproot and a solid, erect stem. The leaves are decompound, lightgreen, alternate, variously pinnately divided, having lanceolate to oval segments. It bears auxiliary and terminal compound umbels of white flowers. The fruits are ribbed, ellipsoid, green to greenishbrown when immature, turning reddish-brown at maturity. The odour of the fruit is slightly characteristic of terebinthinate and becomes strong on crushing. The taste is slightly bitter and extremely pungent. The chromosome number is 2n = 11.

Varieties

Only two varieties: 'Suttons and Monoica' are reported in this crop.

Soil

It can be grown on a variety of soils, but thrives best in well-drained sandy loam to clay loam soils containing a moderate amount of organic matter. Any soil which is saline, alkaline or water-logged is unsuitable for its cultivation.

Climate

The plant requires a mild, cool climate in the early stages of growth, and warm and dry weather at maturity. Recently, its successful cultivation in some subtropical parts of the country has also been reported. It does not like heavy rainfall during its growing period and under such conditions the yields are considerably lower.

Cultivation

The crop is propagated by seeds. They can be either sown directly into the main field or by transplanting the seedlings. Since the directly seeded crop takes a longer time in the field, involving higher inputs on irrigation and weeding, the crop is usually raised by transplanting the seedlings.

(a) Nursery Raising and Planting

The seedlings are raised in well-prepared nursery beds of 1×3 m in size. Before sowing, a mixture of equal quantities of calcium amonium nitrate and single super phosphate at the rate of 150 g per bed is added and mixed well with the surface soil. About 1.5 to 2.0 kg of seeds will be enough to raise seedlings for 1 ha of land. After sowing, the seeds are covered with a mixture of FYM and soil, mulched and watered. The beds are kept moist by light irrigations every day. The seeds start germinating after 7–8 days of sowing. At this stage, the mulch is removed and the beds are kept free of weeds and watered continuously. The seedlings are ready for transplanting after about 8 weeks. For transplanting, the seedlings are carefully removed to avoid any damage to the roots. They are transplanted in well-prepared land at a spacing of 45×30 cm or 45×60 cm in the main field.

(b) Direct Sowing

However, where the area is very large and there is a scarcity of labour, direct sowing may be preferred. A seed rate of 12 kg/ha is recommended when the crop is to be raised by direct seeding. Since the seed is very small, it is mixed with fine soil or sand in a ratio of 1:10. The seeds can be sown in furrows at 45 cm apart, from August to the middle September. If there are no rains within a week of sowing, the field should be irrigated liberally. The seed starts germinating after a fore night and it takes about a month to complete the germination for the whole field. When the seedlings develop true leaves and are 8-10 cm in height, thinning is done so that plant-to-plant distance m a row is maintained at 30 cm.

Manures and Fertilizers

The crop has been found to respond well to the application of manures and fertilizers. A fertilizers dose of 100 kg N, 115 kg P_2O_5 and 30 kg K_2O/ha has been reported as optimum for the crop. This is an addition to 20–25 tonnes of FYM. The manure and half of the N, the full dose of P and K are applied in furrows before transplanting. The rest of the N is given as a top-dressing after thinning, when the plants have started branching.

Manures and Fertilizers

The crop has been found to respond well to the application of manures and fertilizers. A fertilizers does of 100 kg N, 115 kg P_2O_5 and 30 kg K_2O/ha has been reported as optimum for the crop. This is in addition to 20–25 tones of FYM. The manure and half of the N, the full does of P and K are applies in furrows before transplanting. The rest of the N is given as a top-dressing after thinning, when the plants have started branching.

Irrigation

Ammi majus is a herbaceous crop and the availability of moisture has a direct effect on its growth and productivity. Hence, it should be irrigated frequently during the dry period. The first irrigation is given immediately after transplanting and afterwards at an interval of 8–10 days. Since the crop is quite sensitive to excess moisture care should be taken to avoid water-logging.

Interculture

As the crop is very sensitive, it is necessary to keep the plots weed-free at the initial stages. Once they grow, they cover the whole plot and thus there is very iittle scope for the other weeds to compete. About 2–3 hand-weddings will be enough to keep the weeds down. In large-scale plantations, two hoeings should be given to keep the soil mellow and free from weeds.

Pests and Diseases

A bright yellow mosaic disease in the form of a ring or line or patterns on the leaves has been reported from Lucknow for the first time on this crop. Several fungi are known to cause damage to the fruits during storage. Aspergillus ochraceus was observed to cause heavy damage to xanthotoxin during the first month of storage, while Fusarium oxysporum causes damage to A. niger and A. flavus and during the first three months of storage and drastically reduces the xanthotoxin content.

Harvesting and Yields

Ammi majus is a 5–6 months' duration crop. When sowing is done during the months of August-September, the crop starts flowering during the months of November-December. The flowering and maturity of seeds is spread out over about a month.

Studies have shown that the mature, brown coloured fruit contains only 0.4% xanthotoxin; whereas, when harvested at the mature green stage, it gave 0.72% and in immature green fruits, it was as high as 1.0%. However, the mature green stage of harvesting is recommended as, in the case of the immature green stage, there is a considerable reduction in the total yield of the crop. The other advantage of harvesting at this stage is that the loss of the crop due to seed-shattering, which occurs when the seeds have turned brown, can be minimized.

While harvesting, the umbels are hand-picked individually by cutting the plant at its middle. The cut umbels are kept in loose bundles and staked till the fruits are dry. The fruits are later separated by threshing and winnowing.

On an average, the crop may yield about 900 to 1 200 kg/ha of dry seeds.

Ipecac

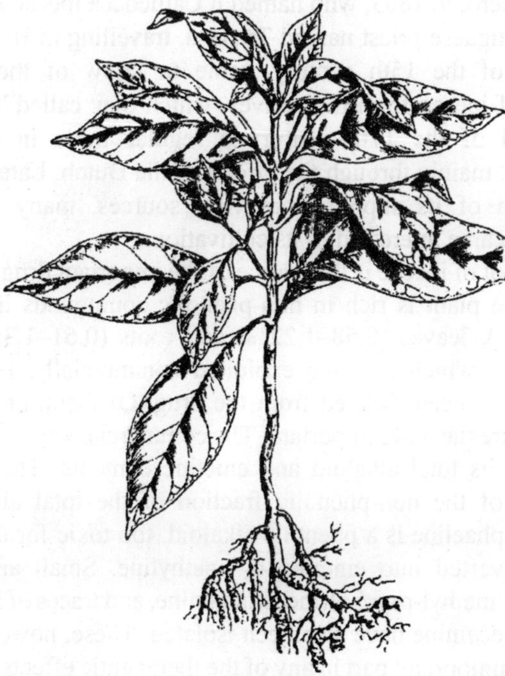

PLANT PROFILE

Family	:	Rubiaceae
English name	:	Ipecac
Species and varieties	:	Cephaelis ipecacuanha C. acuminata Karst. Rio or Brazilian and Cartegena
Distribution	:	India, Bolivia, Brazil, Burma
Uses	:	Drugs

Ipecac Cephaelis ipecacuanha (Brotero) A. Richard is one of the oldest medicinal plants known to man. It IS said to have been used by South American Indians before the beginning of recorded history. Dr. Adrian Helveties of Paris is credited with the introduction of ipecac to Europe. The early Brazilians appear to have applied the term ipecacuanha to a number of roots with common emetic properties, until a Portugese scientist, Berhardino Gomes, collected the authentic plants. These plants were later described by Flex de Aucllor Brotero, in 1803, who named it Callicocca ipecacuanha.

A Portuguese priest named Trislasu, travelling in Brazil in the later years of the 15th century, came to know of the curative properties of ipecac from the natives, which they called 'Igpecaya'. By the end of the 17th century a regular trade in this drug commenced, mainly through the efforts of the Dutch. Later, for fear of depletion of its supply from wild sources, many European countries became interested in its cultivation.

The root of ipecac is the source of the important drug 'emetine'. Although the plant is rich in non-phenolic compounds in its stem (0.21–0.54%), leaves (0.58–1.22%) and roots (0.61–1.35%), it is only the root which is being exploited commercially. Five major alkaloids have been isolated from the drug. Of them, emetine and cephaeline are the more important. The commercial value of the drug depends on its total alkaloid and emetine contents. The principal constituent of the non-phenolic fraction of the total alkaloids is emetine. Cephaeline is a phenolic alkaloid, too toxic for use, and is usually converted into emetine by methyline. Small amounts of psychotrine, methyl-psychotrine, emetamine, and traces of ipecamine and hydroipecamine have also been isolated. These, however, playa relatively unimportant part in any of the therapeutic effects of ipecac. The principal therapeutic use of emetine is in the treatment of amoebiasis. Besides the alkaloids, the roots contain choline, a colourless glycoside (ipecaccanhin), a phytosterol (m.p., 160–162) and a crystalline colouring matter, erythrocephaline. The presence of several glycosides, tannins, malic and citric acids, saponin, resin, fat and a large amount of starch in the roots has been reported.

Ipecac is an important drug being at once an emetic, a diaphoretic, one of the convenient known expectorants, and an excellent remedy for the treatment of amoebic dysentry. A dose smaller than 0.03 g of the dry drug is a stimulant to the stomach, exciting appetite and facilitating digestion. It is used in the form of syrup, powder, tinctures and lozenges. It has become one of the most prominent chemicals used in cancer research. A mixture of ipecac powder and opium was prescribed as a sudontic at the onset of influenza, but this is no longer official. The syrup of ipecac is often given in the early stages of acute bronchitis. It has been used successfully in the treatment of bilharziasis, guinea-worms and oriental sores. The chief adulterant of the drug is the aerial stem of the plant. It can be distinguished from the root by the pith composed of cells with lignified walls and by the surface scars.

When administered in large doses (0.03–0.12 g) it is nauseating, a diaphoretic and expectorant. It acts as an emetic resulting in myositis of the injection site, diarrhoea and vomiting, neuromuscular sympt6ms, hypertension and heart palpitations and also circulatory failure. Powdered ipecac is a respiratory irritant causing coughing and sneezing. Pharmacists and assistants scentitized by repeated exposure may experience rhinitis and asthma.

Origin and Distribution

This plant is a native of the humid forests of Bolivia and Brazil. It has been grown on a commercial scale in Malaysia, and to a small extent in Burma, Brazil and India. It is cultivated in Mungpoo, Rongo, Munsong and Latpanchor (Darjeeling District) as a subsidiary crop in cinchona plantations and in the Niligiris, especially near Kallar. It has been successfully raised at the Rungbee Cinchona Plantation in Sikkim. Jorhat and Srimargal in Assam, Sinduhans and Jalpaiguri in West Bengal and Bilaspur in Orissa are also considered to be suitable localities for its cultivation. The total area under this crop in India is about 100 ha.

Description of the Plant

The species ipecacuanha has been placed in several genera. However, the genus Cephaelis has been considered as correct, according to the International Code of Botanical Nomenclature. This is the name recognized by the US and British pharmacopoeias. The plant belongs to the family Rubiaceae and has the chromosome number 2n = 22.

It is a tropical perennial, herbaceous plant, 50–75 cm high, with a slender, cylindrical, corky, sinuate, underground stem (or false rhiozme) from which extend several horizontal roots. When mature, the roots have a brick-red to dark-brown bark, smooth or more or less prominently transversely ridged or ribbed. They possess a musty odour and have a bitter taste. The stem above ground-level is quadrangular, somewhat downy, with few or no branches. The leaves are Opposite near the top of plant, alternate below, dark-green on the upper surface and light-green below, rough, hairy with short, downy petioles. The flowers are white, funnel-shaped, sessile, less than 1 cm wide and are borne in dense, globular clusters of 8–20 on a purplish-green stalk. The fruit is oval, purple, becoming nearly black when ripe and contains two hemispherical stones, each enclosing a single white seed.

The rhizome is slender, bearing horizontally-spreading fibrous roots. The latter are smooth, slender and whitish when young and develop, on maturation, into a thick, brownish bark with numerous closely placed transverse furrows, giving the roots an annulated appearance. These constitute the drug of commerce.

Types of Varieties

Two types of ipecac, Rio or Brazilian and Cartegenian, are recognized by the pharmacopeias of the US, UK and India. The Rio or Brazilian ipecac is obtained form Cephaelis ipecacuanha.

Cartegena ipecac (Syn. Nicaragua, Savanilla and Panama ipecac) is obtained from C. acuminata Karst which grows in the moist forests of Columbia. The plant is similar to the preceding one but bears white fruits. The roots are greyish-brown and less distinctly ridged.

There are two varieties of cartagena ipecac, one is a greyishbrown form; the other is reddish-brown, comes largely from Nicaragua and is appropriately called Nicaraguan ipecac. This variety is characterized by numerous transverse ridges on the roots that frequently bear light-coloured abrasions.

Many strains of ipecac exist and several commercial varieties have been distinguished, the more important among them being known by names indicating the country of origin. The Rio, Matto Grosso, Minas and Manaos varieties, all come from Brazil. Indian and

Johore ipecac are from Bengal, Burma and the Federation of Malaya. The best commercial drug comes from Matto Grasso in Brazil and is collected from wild plants. Two strains of C. ipecacuanha exist at present in Brazil, one of which appears to be the source of the commercial Matto Grasso drug and the other of the Johore drug.

Soil

The plant requires deep soil with a high humus content. It thrives best in forest areas in sandy loam soils, rich in potash, magnesium and lime. Heavy soils and water-logged conditions, which provide poor aeration for the roots, are not suitable for growing this crop. The ideal soil pH is reported to be around 6.

Climate

Ipecac is a shade-loving plant, sensitive to direct exposure to sun and light, and prefers high humidity and moisture and also protection from winds for its growth and development. It thrives well in a tropical mild, humid climate. An annual rainfall of 150–500 cm well distributed throughout the year and a temperature ranging from a minimum of 18-20°C to a maximum of 36–38°C, are essential for its successful cultivation. Temperature fluctuations should not be great. The plant is sensitive to frost, but can be grown successfully at altitudes of 350–850 m.

Cultivation

Propagation

The crop is raised from seeds as well as cuttings from old plants, petioles and adventitious buds produced by roots. The vegetative method of propagation is not favoured, because the root-yield is poor. Commercially, the plant is propagated by seeds.

In seed propagation, it can be raised either by direct sowing or transplanting seedlings grown in nursery beds.

Nursery Raising and Transplanting

The seed-bed and nursery preparation require very thorough soil working, the removing of all weeds, roots and stones from the soil. Terraces are cut off at 3 to 4 m in width and the soil is dug to a depth of about 3.5 m and thoroughly mulched. The soil is mixed with leafmould and organic manure and the terraces are levelled, so as to give a slight slope downwards to drain off excess irrigation and rain water. The nursery beds are covered with thatched roofing raised with the help of bamboo poles.

In some places, strong polythene sheets instead of thatched roofing are also used. About one-third of the seeds are sterile and so. the seeds are sown @ 35 g/sq m. Before sowing, the seeds are treated with lime for 48 hrs, or with hydrogen peroxide for 96 hours. In the nursery, the seeds are either broadcasted uniformly or, preferably, sown thickly in lines about 5 cm apart. The sowing is done between January and March, depending upon the local climatic conditions. The germination of seeds is slow and erratic, it takes between 3 and 5 months. The small (up to 4 cm height), seedlings are transferred to another nursery-bed at a spacing of 5×5 cm. This is done during August-September, when the monsoon rains are over. The young plants are allowed to remain in this nursery till March-April of the following year. The young growing plants are then transplanted into the main field at a spacing of 10×10 cm or 15×15 cm and allowed to remain there for the next 3 years, till they are dug out upon maturity. Growing the plants under the shade of plants like Teprasia candida and Cratalaria anagryoides and also inter-cropping with Cinchona ledgeriana have given encouraging results.

Direct Sowing

In direct sowing, the seeds are either sown at a spacing of 15×15 cm or sometimes also broadcasted.

Manures and Fertilizers

The use of organic manures, leaf-mould and compost has been recommended to increase the quantity of nutrients in the soil and to improve drainage. There is not much information available on the nutritional requirement of ipecac. In an experiment conducted at Mungapoo, the highest root yield has been reported by the application of castor-cake @ 600 kgl; whereas, superphosphate in combination with nitrogenous fertilizers, augmented the formation of non-phenolic alkaloids.

Irrigation

The nursery-beds require very careful irrigation. A watering can with a fine, low nose is used to water the seed beds in order to keep the soil moist and avoid crust formation. The crop in the field is also provided with regular irrigation during the dry period.

Weeding

A schedule of weeding-cum-hoeing should be followed in such a way that the plantation is kept free of weeds and the soil remains porous. Mulching the nursery beds with coloured polythene sheets (25 gauge in thickness) has been reported to reduce the cost of weeding. The sheet is perforated at the places where the seedlings are transplanted.

Pests and Diseases

The plants are affected by damping-off fungi like Rhizoctonia spp. in the nursery. As a control measure, the seeds should be treated with a suitable seed-dressing of fungicide before sowing. Fusarium moniliforme is reported to cause wilt disease, for which suitable control measures like field sanitation and drenching the plants with fungicidal solution like copper oxychloride (3 g/l) may be taken up. The other diseases reported on this crop are leaf-blight and rot caused by the fungi, Alternaria alternata and Fusarium spp., respectively.

Harvesting and Processing

The plant is a perennial herb. One-fourth of the planted area is uprooted and replanted annually, to maintain a uniform market supply. It survives for up to 5–6 years. It starts flowering from the second year onwards, and its fruits take 3 months to ripen. The production of fruits peaks during the 3rd to 4th year. The formation of alkaloids increases with the age of the plants and roots and is the most during the 3rd-4th years. Therefore, harvesting should be done after the plant is 3 years old.

At the time of harvesting, the roots are dug out from the soil cleaned of the adhering soil and exposed to the sun in a thin layer for drying, as quickly as possible. The roots are kept covered during the night to protect them against heavy dew. The drying takes 2–3 days, thereafter the roots are broken into small pieces and sieved to remove the remaining adherent soil. The root pieces, finally packed into bales, are ready for export or transported to extraction plants for the production of emetine alkaloids and other preparations.

Yields

A 3-year-old crop (4 years from seeds) yields the maximum amount of roots and also contains a high alkaloid content. The Indian Pharmacopoeia requires that the roots contain not less than 2% of the total alkaloids, of which not less than 50% should be emetine in content.

A 3-year-old plantation, on an average, may yield up to 6.5 t/ha of fresh roots, when managed properly.

Isabgol

Plant Profile

Family	:	Plantaginaceae
English name	:	Blond psyllium, Spogel seeds
Indian name	:	Snigdhabijah, Snigdhajirakah (Sanskrit) Isabgol, Isabgul (Hindi) Iskol, Isphogol (Tamil)
Species and varieties	:	Plantago ovata Forsk P. Psyllium linn.

		P. insularis Eastw.
		GI-l, GI-2, Niharika
Distribution	:	India, West Asia, West Pakistan, Persia Mexico, Mediterranean Regions
Uses	:	Drugs

Plantago is a large genus of herbs and subshrubs, distributed mostly in the temperate regions and a few in the tropics. It comprises of about 200 species, of which 10–14 are natives of India. Plantago ouata Forsk (Syn. P. isphagula Roxb.), commonly known as Isabgol or blonde psyllium or Indian plantago or psillium in commerce, is important for its seeds and husks which have been used in indigenous medicine for many centuries. Because of the superior quality and yield of its seeds, it has replaced the French psyllium, P. psyllium Linn, in the world market. In fact, France now depends on India for its requirements.

It derives its name from two Persian words, 'asp' and 'ghol' meaning a horse-ear, referring to its characteristic boat-shaped seeds. Plantago is from Latin, meaning 'sole of the foot' and refers to the shape of the leaf, Psyllium is from the Greek, meaning 'flea,' in reference to the colour, size and shape of seeds (flea-seed) and ovata refers to the ovate shape of the leaves.

The husk of the seed is the economic part and is separated by a physical process. The seed-husk contains a colloidal mucilage (30%), mainly consisting of xylose, arabinose, galacturonic acid with rhamnose and galactose, etc. The seed also contains some oil and small amounts of glycoside acubin and tannin. The husk has the property of absorbing and retaining water and, therefore, it works as an antidiarrhoeal drug. It is beneficial in chronic dysentries of amoebic and bacillary origin. It is also used for treating constipation and intestinal disorders because it works as calorie-free fibre food, promoting regular bowel movement. It is reported to have no adverse side-effects.

The seed has also cooling and demulscent effects and is used in Ayurvedic, Unani and Allopathic systems of medicine. The seeds and husks are used to cure inflammations of the mucous membranes of gastro-intestinal and genitourinary tracts, duodenal ulcers, gonorrhoea and piles. It can also be used as a cervical dilator for the termination of pregnancy. In addition to these medicinal uses, it has a place in dyeing, calico printing, in the ice-cream industry as a stabilizer, also in confectionery and cosmetic industries. The seed without the husk, which contains about 17–19% protein, is used as cattle-feed.

Distribution, Area and Production

The plant is indigenous to Persia and West Asia, extending up to the Sutlej, Sind and West Pakistan. The plant has also acclimatized well in Mexico and the Mediterranean regions.

At present, the Isabgol crop has acquired the place of the 'dollar earner' crop of north Gujarat, where it is being cultivated on about 50000 ha with a production of 40000 tonnes. It has now been introduced in Southern Rajasthan, Punjab and, to a very small extent, in Maharastra and Uttar Pradesh. Experimental cultivation at Bangalore has been quite successful. The quality of husk and yield is comparable to the crop grown in Gujarat. About 75–80% of the total annual produce of north Gujarat is exported, which brings crores of rupees of foreign exchange annually. India continues to hold a monopoly in its production and trade in the world. Though India enjoys the monopoly in production and export of Isabgol, hardly 50% of the requirement of the USA is being met. In view of its s ustained demand, there is still scope to increase the area and production and to intensify its cultivation. Therefore, if India is to retain its monopoly in the production of this important foreign exchange earning commodity, there is an urgent need to evolve high-yielding strains and to examine suitable locations for growing this crop in other states of the country.

Description of the Plant

P ouata (2n = 8), belonging to the Plantaginaceae family, is a stemless or short-stemmed, highly cross-pollinated, annual herb which attains a height of 30 to 40 cm, has alternate leaves, which clasp the stem, strap-like, recurved, 7.5 to 2.5cm long, narrow-varying from less than 6.0 mm to 12.0 mm in width, tapering to a point, three-nerved, entire or toothed, coated with fine hairs. The flowers are white, minute, four-parted. The capsule is ovate, 8.0 mm long, 2-celled, the top half lifting up when ripe, releasing the smooth, dull, ovate seeds which are 1.8 to 3.8 mm long, pinkish-grey brown or pinkish-white with a brown streak on the convex surface. The seeds are covered with a translucent membrane, known as the husk, which is odourless and tasteless. When soaked in water, the whole seed appears hugely swollen because of the expansion of the mucilage in the husk. The husked seeds are dark-red and hard.

Another species, P insularis Eastw. is closely related to P ouata and is different from it in having rachis villous, the trichomes halfas-long to longer than the bracts, the bracts being pubescent, rarely glabrous, the membranous wing of the basal bract is as wide as or wider at the apex than at the base. The filaments and style are red, pink or white, the corolla mid-vein is often reddish.

Varieties

Gujarat Isabgol-l (GI-l) and Gujarat Isabgol-2 (GI-2) are the two varieties of this crop released by the Gujarat Agricultural University. Both these varieties were identified, evaluated and recommended for cultivation based on their yield performance alone, and none of these varieties have any markers by which these can be distinguished. Another variety, 'Niharika', has been released by the CIMAP, Lucknow, as a high yielding variety.

Soil

It is an irrigated crop which grows well on light soils. Heavy soils and those with poor drainage are not conducive to the good growth of this crop. A silty-loam soil, having a soil pH from 4.7 to 7.7 with low nitrogen and moisture content, is reported to be ideal for the rich growth of these plants and high yield of the seeds.

Climate

Isabgol thrives in warm temperate regions. It requires cool, dry weather and is sown during the winter months. Sowing during the first week of November gives the best yields. Early sowing, especially with a higher seed rate, makes the crop vulnerable to downy mildew diseases; whereas, late sowing provides a shorter period of growth in winter along with the possibility of seed-shattering due to summer rains in April-May. At maturity, if the weather is humid, its seeds shatter, resulting in reduced yield. Heavy dew or even a light shower will proportionately decrease the yield, sometimes even leading to a total loss of the crop. The temperature required for maximum seed germination is reported to be between 20° and 30°C.

Cultivation

Land Preparation

The field must be free of weeds and clods and should have a fine tilth for good germination. The number of ploughings, harrowings and hoeings depend upon the soil conditions, the previous crop and the degree of weed infestation. About 10-15 t of FYM/ha is incorporated into the soil at the time of the last ploughing. The field should be divided into suitable plots of convenient size, depending upon the texture of the soil, the slope of the field and the source and quantum of irrigation water. For light soils with even contours, a plot size of 8.0 m × 3.0 m is convenient.

Seeds and Sowing

For a high percentage of germination, the seeds should be taken from the crop harvested at the end of the preceding season. Older seeds tend to lose viability under ordinary storage conditions. The seeds at the rate of 4–8 kg/ha are sown after treating them with any mercurial seed-dresser at the rate of 3 g/kg seed, to protect the seedlings from a possible attack of damping-off.

The seeds are small and light. Hence, before sowing, the seed must be mixed with a sufficient quantity of fine sand or sieved FYM. The seeds are broadcast, because sowing in lines at different spacings does not increase the seed yield significantly. After broadcasting, the seeds are swept over lightly with a broom to cover them with some soil. The broom, however, should be swept in one direction only, to avoid deep burial of the seed and for uniform germination. The sowing should immediately be followed by irrigation. Germination begins four days after sowing. If delayed, it should be stimulated by another watering.

Fertilizers Application

Isabgol does not require the application of heavy doses of fertilizers. A fertilizer dose consisting of 50 kg N, 25 kg P_2O_5 and 30 kg K_2O/ha, has given the maximum seed yield under Bangalore conditions. The full dose of P and K along with half of the N is given as a basal dose at the time of sowing itself, and the second split dose of N is applied as a top-dressing after one month of sowing. For other areas, a basal dose of 25 kg N and 25 kg pp/ha is recommended. A top-dressing of 25 kg N/ha is given 32 days after sowing.

Irrigation

Immediately after sowing, light irrigation is essential. First, irrigation should be given with a light flow of water otherwise, with a fast current of water, most of the seeds will be swept off to one side of the plot and the germination and distribution will not be uniform. The seeds germinate in 6–7 days. If the germination is poor, a second irrigation should be given. Later on, irrigations are given as and when required. The last irrigation should be given at the time when the maximum number of spikes have reached the milk-stage. The crop requires totally 6–7 irrigations for good productivity in medium sandy soils.

Interculture

The first weeding is done after about 20–25 days of sowing. Since the crop is sown by broadcasting, hand-weeding is a very costly operation. Ordinarily, 2–3 weedings within two months of sowing will help to keep the weeds under control.

Pests

Insect Pests

Sometimes, white grubs and termites damage the crop by cutting off the root. These can be controlled by broadcasting phorate 10G.@10 kg/ha before the last ploughing. Aphids also attack this crop, and can be controlled by spraying 0.2% Dimethoate.

Diseases

Downy Mildew

It is a major disease of this crop which is caused by Peronospora plantaginis and P. alta. The extent of the damage depends upon the intensity of the disease. Normally, this disease appears at the time of spike-initiation. The first symptom is small patches on the leaf, which spread throughout the leaf, completely destroying it and thus affecting the yield. The disease intensity will be very high in humid weather or when the dewfall is high. It is very difficult to keep the disease under control once it has affected the crop extensively. Hence, as a prophylactic measure, it is suggested to spray one of the following fungicides whenever the weather becomes cloudy and humid.

1. Bordeaux mixture (6:3:100) at the rate of 600 to 800 Vha.
2. Copper oxychloride, Dithane M-45, Dithane 2–78 or any other copper fungicide at the rate of 2.0 to 2.5 g in 1 litre of water.
3. The antifungal antibiotic, Aureogungin. The fungicide should be sprayed 30 days after sowing, followed by two or more sprays at an interval of 15–20 days.

The crop is also sometimes affected by the damping-off disease, caused by Pythium ultimum, Rhizoctonia solani and Fusarium oxysporum. Wilt is reported to be incited by Pultimum, Foxysporum and Fsolani. Another disease of minor importance, the powdery mildew which appears at flowering time, is caused by Erysiphe cichoracearum.

Harvesting and Processing

Blooming begins two months after sowing and the crop is ready for harvest in February-March (110–130 days after sowing). When nature, the crop turns yellowish in colour and the spikes turn brown in colour. The seeds are shed when the spikes are pressed even slightly. At the time of harvest, the atmosphere must be dry and there should be no moisture on the plant, otherwise harvesting will lead to considerable seed-shattering. Hence, the crop should harvested after 10 a.m.

The plants are normally cut at the ground level or are uprooted, depending upon the texture of the soil. The harvested plants are transported to the threshing yard, bundled in large cloth pieces and spread out. After 2 days, they are threshed with the help of a tractor or bullocks. Threshing should preferably be done during the early morning hours to facilitate the easy separation of the seed from the spike. Sometimes, water is sprinkled over the heap for easy threshing. Care should be taken not to sprinkle excess water, lest it should spoil the quality of the produce. Its straw is used as cattle feed. The plants are threshed and winnowed and the seeds sieved until they are clean. The seeds may be marketed whole or the husk may be sold separately. To remove the husk, the cleaned seeds are passed 6 to 7 times through stone-grinders, sieved and screened through several grades of mesh to sort according to fineness. Most of the exported husk is of the grade designated as 70 mesh. The highest quality husk is white with no particles of the red kernels. The husk seed ratio is 25:75 by weight. It costs about Rs. 2–8 million (at 1978 prices) to establish an industry for processing about 8 t of seeds per day.

Yields

Gujarat Isabgol 1 variety yields 800–900 kg of seeds/ha. The new variety, 'Gujarat Isabgol 2' has a potential to yield a 1000 kg of seeds/ha. Increased yield can be obtained in this crop by spraying cycocel (CCC) at 25 mg/l. twice: at the seedling stage and just before the heading stage.

Analytical Methods

The isabgol husk is used for its mucilage content, mucin or mucilage are generally acidic mucopolysaccharides. They are usually attached to proteins to form proteoglycans, in which the polysaccharides make up most of the weight. Normally, mucin consists of recurring disaccharide units each of which contains a derivative of an aminohexose, usually D-glucosamine or D-galactosamine. The polysaccharides isolated from mucin are slippery and lubricating proteoglycons of mucous. These are jelly-like ground substance or intercellular cement. There is scanty information available regarding the composition and quality of the mucilage of isabgol. For screening of isabgol germplasm, it is necessary to test the isabgol mucilage for its quantity and quality parameters. To study the swelling capacity of the husk, 1g of husk is put in a 50 ml centrifuge tube and 40 ml water is added to it. After mixing the contents, the tube is kept overnight. The next day, the tube is centrifuged for 5 minutes at 5 000 R.PM. When the gel settles down and a clear liquid is separated on top, the volume of gel is measured, that is 34 ml, which shows the swelling capacity to be 1 :34.

Kalmegh

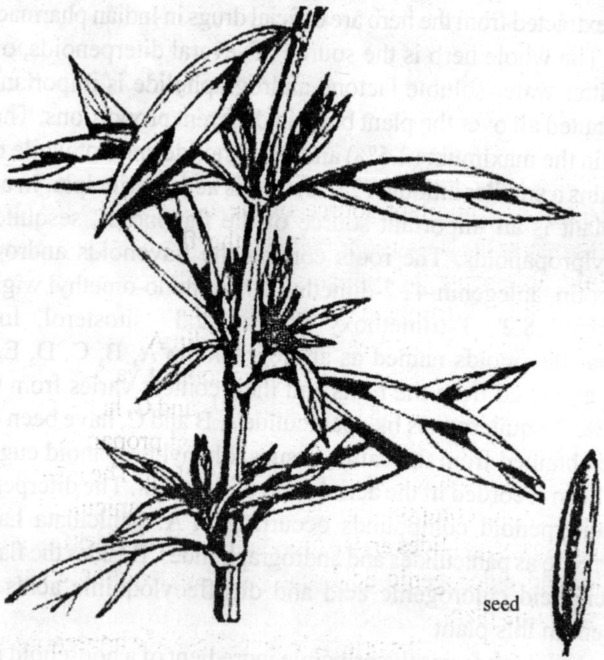

Plant Profile

Family	:	Acanthaceae
English name	:	King of Bitters, Chiretta
Indian name	:	Bhunimbah, Kirta (Sanskrit), Kalmegh (Hindi) Nelaberu (Kannada), Nilavembu (Tamil)
Species and varieties	:	Andrographis paniculata Nees.
Distribution	:	India, Bangladesh, China Sri Lanka, Thailand

Uses : Drugs

Kalmegh (Andrographis paniculata Nees.), belonging to the family Acantaceae, is one of the nineteen species of the genus Andrographis, which is indigenous to India and has been used in Indian systems of medicines since time immemorial. The plant is also known as 'Rice bitters' in West Indies and 'King of bitters' or 'Chiretta' in England. The fresh and dried leaves of kalmegh and the juice extracted from the herb are official drugs in Indian pharmacopoeia.

The whole herb is the source of several diterpenoids, of which the bitter water-soluble lactone andrographolide is important and is distributed all over the plant body in different proportions. The leaves contain the maximum (2.5%) andrographolide content while the stem contains a smaller amount (2.0%) of this active principle. In addition, the plant is an important source of the flavonoids, sesquiterpenes, phenylpropanoids. The roots contain the flavonoids andrographin, panicotin, aplegenin-4', 7-dimethyl ether, mono-omethyl wightin and hydroxy-7,8,2', 3'-trimethoxy flavone and sitosterol. In Japan, several flavonoids named as andrographidin A, B, C, D, E, F have been extracted from the roots and their content varies from 0.015 to 0.15%. Sesquiterpenes like paniculide A,B and C, have been reported to be obtained from the callus tissues. Phenyl propanoid eugenol has also been recorded in the aerial parts of the plant. The diterpenoid and sesquiterpenoid compounds occurring in A. paniculata have been referred to as paniculides and andrographolides. Besides the flavonoids, caffeic acid chlorogenic acid and dicaffeoyloquinic acids are also present in this plant.

Kalmegh forms the principle ingredient of a household medicine called 'Alui', extensively used in West Bengal for general debility and for certain forms of dyspepsia amongst adults and infants. The expressed juice of the leaves is prescribed with cardamom, cloves and cinnamon in the form of globules to infants for relief from bowel complaints, irregular stools and loss of appetite. The plant is considered to be highly efficacious against chronic malaria and often used as a substitute for Swertia chirata. In India, the entire plant is used to treat snakebite. The hot water extract of the whole plant is used for acute jaundice, where the powder is mixed with garlic and 2g of this mixture is given orally with buttermilk for four days and also as febrifuge and as an antidysentric agent; while the extract of the dried leaf is used to treat stomach worms. The hot water extracts of the dried leaf and stem is used as a stomachic, febrifuge, and tonic, while the fresh leaf-juice along with the leaf-juice of Azadirachta indica and Tinospora cordilolia are used to cure cholera. The roots are used as antipyretic, alternative and cholagogue agents. The decoction of the plant is used in some cases of anemia. In Thailand, the decoction of the dried leaf is used against high blood pressure. While in Bangladesh, the hot water extracts of leaf and stem are prescribed as a powerful tonic.

The common ayurvedic drugs are 'Kalmeghasva' and 'Kalmeghnamayas Haub', the main constituents of which are A. paniculata besides Trikatu (Piper longum, P. nigrum and Zingiber officinalis), Cyperus scariosus, Triphala, Embelia ribes, Plumbago zeylanica and Lauh Bhasm (organo-metallic salt of iron).

Distribution

It grows wild as an undershrub in tropical, moist, deciduous forests. The plant is found in the plains throughout India and Sri Lanka. It is also reported from certain parts of China, Thailand and Bangladesh. In India, it is distributed in the States of Andhra Pradesh Assam Bihar, Karnataka, Kerala, Madhya Pradesh and West Bengal.

Botany

It is an annual herbaceous or shrubby recumbent or erect plant, 0.3–1.0 m high. The branches are sharply quadrangular, often narrowly winged towards the apical region. The leaves are petiolate, 5–8 cm long and 1–1.25 cm broad, lanceolate and acute. The flowers are small, solitary, in panicles with a rose-coloured corolla which is hairly externally. The filaments are flattened, hairy on the upper part and the anthers borne at the base. The fruit is 20 mm long and 3 mm broad, linear-oblong and acute at both ends. The seeds are numerous, yellowish-brown and glabrous.

Soil

Kalmegh is a hardy plant and can be grown on a variety of soils. In the native state, it is found growing in clayey to sandy soils in various locations. However, sandy loamy soil rich in organic matter is good for its growth and yield. Soils with water-logged conditions and problematic soils are not suited for its cultivation.

Climate

This crop comes up well in tropical and subtropical regions. However, cooler climates with well distributed rainfall are ideal. Under natural conditions, this crop is found growing abundantly in moist, shady places and sometimes in dry forests also. Studies at Bangalore have shown that the best season for planting this crop is during the month of June when it has the maximum growth and gives the highest yield of herb.

Propagation

Kalmegh is propagated either by seeds or cuttings. However, for commercial purposes, the crop is raised from seeds as it is not only easy but economical compared to the vegetatively grown crop.

Seed Propagation

Seeds are either sown directly in the main field or nursery raised seedlings are transplanted. About 400 g/ha of seeds are required for sowing. While, for raising the seedlings, the seeds are sown in polythene bags filled with a mixture of soil, sand and organic manure in 1: 1: 1 proportion. The seedlings can also be raised in nursery beds of 3 × 1.5 × 0.15 m size. The nursery beds are mixed with well rotted, powdered cattle manure @ 20 kg/m^2. The seeds are sown in lines 5 cm apart in the nursery beds. The beds or polythene bags are watered daily. The seeds germinate in about 8–10 days and the seedlings will be ready for transplanting into the main field when they are 45–50 days old.

Land Preparation

The land is prepared well by repeated ploughing and planking and brought to a fine tilth. FYM @ 25 t/ha is mixed well into the soil at the time of the last ploughing. The land is laid out into plots of convenient sizes along with irrigation channels for easy management.

Sowing/Transplanting

For direct sowing, furrows are made at a distance of 30 cm and about 3–4 seeds are sown in each spot maintaining a distance of 15 cm between two plants; while the seedlings are transplanted at a spacing of 30 × 15 cm or 15 × 15 cm. Studies have shown that in order to get the maximum herb yield, a spacing of 15 × 15 cm is the best.

Application of Manures and Fertilizers

Kalmegh responds very well to the application of manures and fertilizers. Besides the application of about 25 tonnes of FYM, a fertilizer dose of 75 kg N, 7 5kg P_2O_5 and 50 kg K_2O is required for a crop of one hectare. Of this, 50% N and the entire dose of P_2O_5 and K_2O is given as basal dose and the remaining 50% N is used as a top-dressing 30 days after sowing or transplanting.

Irrigation

As kalmegh is mainly grown as a rainy season crop, in the areas with well distributed rainfall hardly any irrigation is given. However, in case of long dry periods, the crop is irrigated initially at 3–4 days' interval and later the interval is increased to one week, depending upon the weather conditions.

Weeding

Kalmegh is a small herbaceous shrub and, therefore, the competition with weeds during its initial stages of growth should be avoided by keep 10g the land free from weeds. The first weeding is done after about 20–30 days of planting. After this, one or two weedings are done after 60 days of sowing to keep the weeds under control during the initial stages of crop growth. Later, weeding follows every harvest.

Pests and Diseases

Kalmegh is a hardy plant and is not attacked by any pests and diseases of a serious nature.

Harvesting and Yield

The crop is ready for the first harvest after about 90 to 120 days of sowing when the plants start flowering. At this stage they are harvested by cutting the plants at the base, leaving about 10–15 cm of the stem for regeneration. The field is given a top-dressing of N and irrigated after every harvest for regeneration. The regenerated plants are ready for harvest in about 60 days of the previous harvest. Totally, 2 to 3 harvests can be made 10 a year. Sometimes the whole plant is also harvested after about 6 months. After harvesting, the plants are dried under shade for 3–4 days before storage.

The average yield is about 2 to 2.5 t/ha of dry herb.

Chemical Evaluation

To estimate the annrographolide content, the coarsely powdered drug (10 g) accurately weighed, is transferred to a Soxhlet extractor and exacted with 200 ml of alcohol for six hours. The extract is concentrated to about 50 ml; 30 ml of distilled water' is added to it.

The solution is extracted with two 25 ml portions of benzene and the benzene layer is rejected. Distilled water (20 ml) is added to the aqueous layer. The combined aqueous solution is then extracted with three 25 ml portions of ethyl acetate, which is sufficient to extract all the bitter principles. The combined ethyl acetate extract is treated with anhydrous sodium sulfate and then filtered through a cotton plug into a 10 ml beaker. The cotton plug is again washed with 5 ml ethyl acetate. The residue collected in the beaker is dried to constant weight on a water-bath and weighed as andrographolide. The values are expressed in per cent andrographolide.

Liquorice

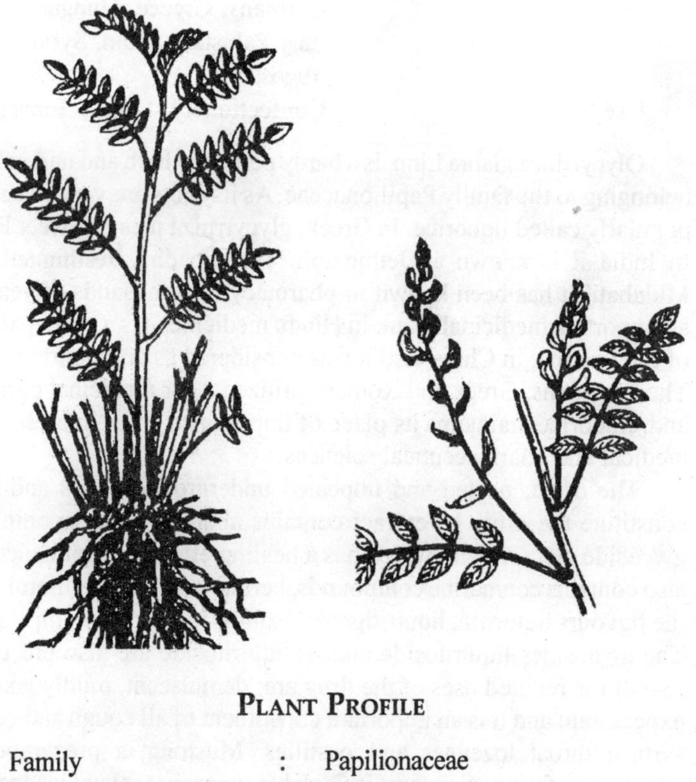

Plant Profile

Family	:	Papilionaceae
English name	:	Liquorice
Indian name	:	Yastimadhu, Madhukah (Sanskrit) Jetimad, Mulahati (Hindi), Atimadhura Yestimadhuka (Kannada), Atimaturam (Tamil)

Species and varieties	:	Glycyrrhiza glabra Linn. typica Regel. and Herd glandulijera. Waldst and Kit violaceae Boiss Haryana Mulhati No.1, EC-21950
Distribution	:	India, Afghanistan, North Africa, Asia Western China, Southern Europe, France Germany, Greece, Hungary, Italy Pakistan, Spain, Syria, Yugoslavia
Uses	:	Confectionery, Drugs, Tobacco

Olycyrrhiza glabra Linn. is a hardy perennial herb and undershrub, belonging to the family Papilionaceae. As its roots are very sweet it is popularly called liquorice. In Greek, glycyrrhiza means 'Sweet Root'. In India it is known as Jethimadh, Yasthimadhu, Jesthimadhu or Mulahati. It has been known in pharmacy for thousands of years as an important medicinal plant. In Hindu medicine, it is a principal drug of the Susruta, in China also it was considered as an important drug. The Egyptians, Greek and Romans utilized it for medicinal purposes and, liquorice maintains its place of importance, even in present day medical and pharmaceutical sciences.

The dried, peeled and unpeeled underground stems and roots constitute the drug. Its extract contains about 5–20% saponin, like glycoside glycyrrhizin, which has a healing effect on peptic ulcers. It also contains coumarine compounds, herniarin and umbelliferone and the flavours liquiritin, liquiritigenin, isoliquirtin and isoliquiritigenin. The glycosides liquiritoside and isoliquiritoside are also present. A few of the reputed uses of the drug are: demulscent, mildly laxative, expectorant and it is an important constituent of all cough and catarrh syrups, throat lozenges and pastilles. Mustong, a preparation in which one of the constituents is G. glabra, improves sexual inadequacy. It is also known for its analgesic, anti-inflammatory and anticonvulsive properties. Recent investigations have shown its growth and inactivation of virus particles. Large quantities of liquorice are used in confectionery and in tobacco blending. The extract is reported to lend sparkle and aroma to beer.

Origin and Distribution

Liquorice is distributed in the subtropical and warm temperate regions, particularly in Western China, parts of Asia Minor, Persia, the Asian Republics of the erstwhile USSR and Afghanistan. It is also distributed in the warmer parts of the Mediterranean region of North Africa, Spain, Italy, Yugoslavia, Greece, Syria, Hungary and South Russia. It is cultivated in Italy, France, Russia, Germany, Spain and China. Although the origin of liquorice is attributed to regions from Southern Europe to Pakistan and Northern India, none of the liquorice yielding species occur in India. In an effort to curb import, substitution, attempts have been made to cultivate liquorice in several places in India, notably Baramulla, Sri nagar, Jammu, Dehradun, Delhi, Anand, Indore and Bangalore and it has been found to come up successfully in quite a few places. Presently, India is importing the drug on a considerable scale (5,000–10,000 t) from Iran, Iraq, Afghanistan, Burma and Singapore.

Description of the Plant

Liquorice (2n = 16) is a perennial herb or undershrub attaining a height of 1-2 m with long, cylindrical, burrowing rootstock and horizontal creeping stolons, which are 1.5–1.8 m in length. The leaves are alternate, pinnate with 9–17 ovate, yellowish-green, leaflets about 2.5–5.0 cm long. The flowers are borne in erect, auxiliary, long spikes, usually 10–15 cm long, lavender to purple in colour and 1.25 cm long. The seed pods are maroon, 3 cm long, oblong, pointed, flattened and contain 2–4 kidney-shaped seeds. The chromosome number is 2n = 16, and the plant is self-pollinated. The root system is highly branched and confined to the upper crust to the soil, it consists of a short tap-root with a large number of rhizomes.

Species and Varieties

1. Glycyrrhiza glabra var. typica Regel. and Herd: This is called Spanish liquorice. It attains a height of 120 to 150 cm and bears typical papilionaceous flowers of a purplish-blue colour. The underground portion consists of long roots and thin rhizomes. The principal root divides just below the ground into several secondaries which penetrate the soil to a depth of 120 cm or

more. A considerable number of stolons are given off, which attain a length of 150 to 180 cm and run near the surface. The concentration of glycocide ranges from 6–10%. It is cultivated in Spain, Italy, England, France. Germany and the USSR It fetches a higher price in the market and is the sweetest of all.

2. Glycyrrhiza glabra var. glandulifera, Waldst and Kit: It is also called Russian liquorice. The root system is shallow, but grows into a large stock; the rhizomes grow over a long distance without stolon development. It is rich in glycyrrhizin (10–14%), but the sweet taste is accompanied by a perceptible but mild bitterness owing to the presence of liquiritin and isoliquiritin.
3. Glycyrrhiza glabra var. vialaceae Boiss: This is called Persian liquorice. It bears violet flowers and is grown in Iran and Iraq, in the valleys of the Tigris and Euphrates. The plant produces thick, hard and lumpy roots and is considered inferior in quality.

Varieties

Through an evaluation trial of 14 accessions of G. glabra from different Countries at Anand. Indore and Haryana, a Russian collection EC-111236 has been released as 'Haryana Mulhati NO.1' by the daryana Agriculture University for cultivation. It has long, broad and ark-green leaves and produces approximately 200 cm-long roots at harvest. This variety is resistant to leaf-spot disease and gives 70-80 q/ha of roots at 3 years with 7.5% glycyrrhizic acid. An accession, EC-21950, has been identified at the All India Co-ordinated Research Project on Medicinal and Aromatic Plants, which is said to be tolerant to the root-rot disease.

Soil

The plant requires a soil with a minimum of 1 m depth for the proper growth and development of the root system. Loamy soils of light texture, with a pH of 6 to 8.2, are ideally suited for high biological and economic productivity. Deep moist soils, particularly on the banks of rivers, subject to periodical inundation, are also suitable. Liquorice is one of the plants which can withstand salinity without affecting its growth and active principles.

Climate

Liquorice prefers a warm and dry, subtropical climate with well-defined winters, to induce the translocation of food to the rhizomes. Heavy rains or frost affects its growth adversely. The plant thrives in locations receiving 50–100 cm rainfall annually, supported with irrigation.

Cultivation

Propagation

Different varieties of Glycyrrhiza glabra are perpetuated by seeds and cuttings of both root and stem and by tissue culture.

Seed Propagation

Though the plants can be grown from seeds, this is seldom practised because of the germination problem. It was found that treatment of seeds with 0.0025 to 0.005% of succinic acid, combined with scarification (in sand), increased laboratory and field germination. The treated seeds are drill-sown 15 cm apart in raised nursery beds (1 × 3m) that are well prepared and manured.

The germination starts after 15 days and is completed within a month. The seedlings are picked out when they are 10–15 cm high and planted in the main field.

Tissues Culture

Rapid clonal multiplication of Glycyrrhiza glabra L., using auxiliary bud culture, has been reported. The technique has also been standardized for transplanting test-tube grown plantlets to the field.

Clonal Propagation

The traditional method of perpetuation is by utilizing the cuttings prepared from the old crowns of the lifted roots, cut into pieces of 10–15 cm length. The plants are also produced from runners or underground stems which are prepared into cuttings of 10 cm long, each with 2 buds. These cuttings are kept in moist sphagnum moss for about 8–10 days and afterwards, when their buds start sprouting, they are planted in the main field.

Preparation of Main Field and Planting

The crop is usually grown continuously on the same land. Uprooting of the plant is done in October-November and this disturbs the soil to a depth of 80 to 100 cm. The soil is then levelled and planting ridges 60 cm wide and 30 cm apart are marked out. The rows are enriched with FYM and the soil from the intervening spaces is swept up to cover the manure to form rounded ridges whose summits are about 37 cm along the furrows. The planting is done in March or early April. The setts are planted in groups of 3, 30 cm apart in rows.

The central planting material of every alternate group is an old crown which is placed at the top of the ridge with the two other setts (root-cuttings) on either side of the ridge approximately 30 cm apart. The crowns are covered with 5 to 7 cm of wet soil. Dry conditions at planting time and for the next few months give the condtions required for a good crop. If cold weather prevails in Mayor June, 20 to 40% of the setts may fail to grow.

Manures and Fertilizers

In order to obtain good growth and development of the roots, the field must be liberally manured. Compost or FYM @ 10-15 t/ha is enough to meet the requirements. However, there is difficulty in getting organic manure in such large quantities, hence, a fertilizer dose at the rate of 40:40:20 kg/ha per year is recommended for application in the form of urea, superphosphate and muriate of potash. Of these, a full dose of superphosphate and potash are applied as a basal dose at the time of planting. Whereas, nitrogen is applied in three split doses: *i.e.*, at planting, after six months and one year of growth, respectively. Since the crop remains in the field for 3–4 years, every year the same dose has to be reapplied.

Irrigation

Regular irrigation is necessary until the cuttings put down roots and establish themselves. Later, the watering may be done at intervals of 8–10 days for 5–6 months and afterwards as and when required. Once the plants are established, they need no special attention as the crop is fairly drought resistant.

Intercropping and Weeding

The crop occupies the land for a period of 4 to 5 years, and the growth is slow during the first 2 years. Thus, intercropping by planting carrots, potatoes or cabbage between the ridges, with the additional inputs required by the intercrop, is recommended. The field may be hoed periodically to keep down the weeds.

Pests and Diseases

Termites cause a lot of damage to young plants during the dry period. Aldrex at the rate of 25–30 kg/ha can be given during the last ploughing/harrowing to control termites. Cotton ash weevil (Myllocerus udecimpunctatus) is reported to cause injury to leaves and this insect can be controlled by spraying Metasystox.

Diseases

The crop is found to be infected by root-rot disease. The roots and stolons of the affected plants turn soft and pulpy; 5 kg/ha Brassicol (or drenching the soil with 6:3:100 Bordeaux mixture) is applied during February to control the disease effectively. Rhizoctonia bataticola, Sclerotium spp.; and Fusarium spp. causes root-rot, collar-rot and wilt diseases. Soil treatment/aerial spray of Bavistin or Benlate (0.05%) is reported to control these diseases.

Cercospora cavarae is reported to cause leaf-spot disease. To control the disease, spray 0.4% Bavistin (4g, 50.w.p. in 10 litres of water). However, one protective spray should be given one and a half to two months after planting. The other fungi which are reported to cause leaf-spot diseases are Alternaria tenuis, Septoria glycyrrhizae or Phyllosticta glycyrrhizae. Uromyces glycyrrhizae causes leaf-rust. Leaf-blight is another serious infection which can be controlled by foliar spraying of Blitox (0.2%) 3 to 4 times, at an interval of 5 days.

Harvesting

The crop is ready for uprooting about 3 to 4 years after planting and just before the plants have borne fruit. The plants are lifted in autumn (November-December) after the rains. A trench about 60 cm deep is first dug at the side of the ridges then, by working inwards, the soil is loosened from the roots so that they can be pulled out easily. The aerial parts are cut and removed. The broken root-parts are left in the soil and new sprouts are thrown out in spring and, for raising the succeeding crop, only the gaps need to be filled with rooted cuttings.

Processing

Roots and underground stems are cut into 15 to 20 cm-long pieces, 1 to 2 cm in diameter and are dried alternately in the shade and sun, this may take several months. The drying process reduces the weight to 50% and the moisture from 50–60%. Artificial drying can also be done at 30–40°C by using mechanical driers.

Yield

The yield depends upon the soil fertility, variety, the climate and cultivation practices. An yield of 5 t/ha of roots, plus 14–20 t/ha of trimmings is considered satisfactory.

Analytical Method

The principal constituent of liquorice is glycyrrhizin which has a characteristic sweet taste. The other constituents are glucose (3.8%), sucrose (6.5%). mannite, starch, asparagine, resins, volatile oil (0.030.06%) and the colouring matter isoliquoricin.

Glycyrrhizin ($C_{42}H_{62}O_{16}$; m.p. 205) occurs as Ca or K salts of glycyrrhizic acid. On acid hydrolysis, it yields, glycyrrhizic acid ($C_{30}H_{46}O_4$, a triterpenoid aglycon) and manuronic acid.

(i) Analytical procedure for *Glycyrrzhiza* Glabra

The shredded roots of liquorice are stirred with 10 times 'its volume of distilled water for 4 to 6 hours, keeping the temperature at 60°C. The pH of the suspension is maintained between 6.5 to 7.0 by adding ammonium hydroxide. The suspension is centrifuged and the supernatant is separated. The extract, thus obtained, is evaporated in vacuum to about 75% of its volume. Ten percent concentrated sulphuric acid solution is added to this suspension with constant stirring. The supernatant liquid is discarded. The precipitate comprising of crude glycyrrhizic acid left in the beaker is taken up in distilled water and the pH of this suspension is brought to 4.0 by adding the required quantity of ammonium hydroxide/sodium carbonate/

potassium carbonate. This suspension is stirred for 4 to 6 hours, and transferred to a centrifuge tube. The suspension is centrifuged till a clear supernatant is obtained. The supernatant is discarded and the residue left behind is suspended in distilled water and its pH is brought to 6.0 by adding ammonia. The material is again centrifuged and the supernatant is discarded. The precipitate is dried and extracted with absolute alcohol. Alcohol is evaporated on a water-bath and dried in a vacuum oven to obtain active glycyrrhizic acid. The recovery of active glycyrrhizic acid under this procedure is 7 to 8%.

(ii) Determination of crude fibre Content in Glycyrrhiza Glabra Roots

The residue left after the water extract is taken and processed as below for the determination of crude fibre content. Otherwise, about 2g of the powdered sample is weighed and treated with ether on a funnel to make it free from fat.

The fat-free residue is transferred to a 400 ml beaker containing 100 ml of 2.5% H_2SO_4 and 100 ml of boiling water, a mark is made on the beaker with a glass pencil to represent the level of the mixture in it. The contents are brought to boil and the boiling is continued for exactly 30 minutes. During the boiling, it is stirred with a glass rod to mix the contents and to remove particles from the sides of the beaker. The volume of the liquid is kept constant by the addition of hot water from time to time. At the end of 30 minutes, the beaker is removed, the contents are filtered immediately through a wet piece of linen cloth and the residue is washed with hot water till it is free from acid.

All the residue from the cloth is then transferred back to the original beaker. The linen is washed with hot water, added to 100 ml of 2.5% NaOH solution and the volume is brought up to 200 ml by the addition of boiling water. The contents are boiled as before for exactly 30 minutes, stirring with a glass rod and keeping the volume constant by the addition of hot water. The beaker is removed and the contents re-filtered immediately through the same linen cloth. All the residue is transferred to the cloth with a jet of hot water. The residue is washed with hot water 3 times, then with dilute 1 % HCl solution and again with hot water till it is free from acid. Finally, the residue is washed once with 95% alcohol and once with ether.

Quantitatively, the whole residue is transferred to a clean dry crucible previously weighed. The residue is dried in the oven at 100°C to constant weight. It is cooled in a dessicator and weighed. The residue is ignited completely, till white or greyish-white ash is obtained. It is cooled in a dessicator and weighed again. The difference in the two weights represents the weight of crude fibre. The glycyrrhiza roots possess about 15 to 20% of crude fibre.

Long Pepper

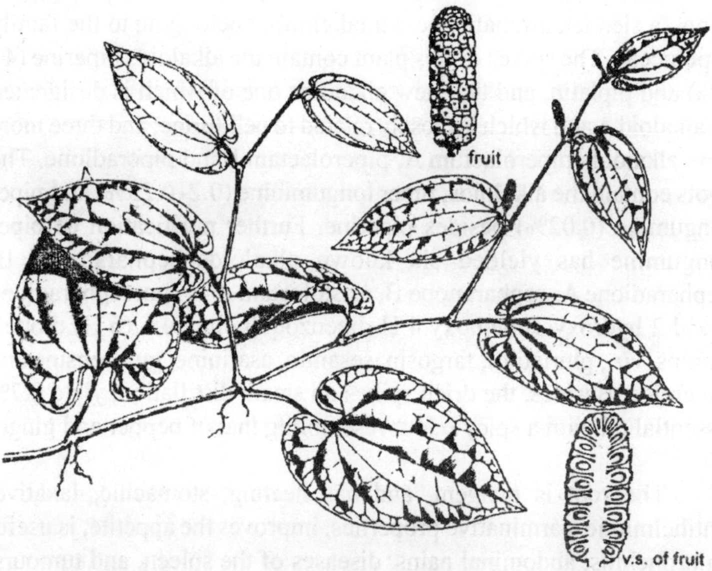

Plant Profile

Family	:	Piperaceae
English name	:	Indian long pepper, Long pepper
Indian name	:	Pippali, Magadhi (Sanskrit) Pipli, Pipal (Hindi), Hippali (Kannada) Pippili, Tippili (Tamil)
Species and varieties	:	Piper longum Linn. P. pepuloides P. officinarum DC.

		Gol Thippali, Pipal Nonsori, Asali, Suvali
Distribution	:	India, Andaman and Nicobar islands Indonesia, Malaysia, Nepal, Philippines Sri Lanka, Timor
Uses	:	Drugs, Spice

Long pepper is obtained from the unripe spikes of Piper longum Linn., a slender, aromatic, perennial climber belonging to the family Piperaceae. The spikes of this plant contain the alkaloids piperine (4–5%) and piplatin, and two new alkaloids one of which is designated as alkaloid A and which is closely related to pellitorine; and three more new alkaloids-piperolactum A, piperolactam Band piporadione. The roots contain the alkaloids, piper longuminine (0.2–0.25%) and piper longumine (0.02%) besides piperine. Further purification of piper longumine has yielded six known alkaloids cepharadione B, cepharadione A, cepharanone B, aristolactam A 11, norcepharadione Band 2 hydroxy 1 methoxy 4 H dibenzoquinoline-4,5 (6 H) dione); lignins, *viz.*, pluriatilol, fargosin, sesamin, asarinine, guineensine and pipercide. Besides, the dried spikes on steam distillation, yield 0.7% essential oil with a spicy odour resembling that of pepper and ginger oil.

The root is pungent, and has heating, stomachic, laxative, antihelmintic, carminative properties, improves the appetite; is useful in bronchitis, abdominal pains, diseases of the spleen, and tumours. The unripe fruit is sweetish, cooling and useful in biliousness. The ripe fruit is sweet, pungent, a stomachic, aphrodisiac, alternative, laxative, antidiarrhoeic, antidysenteric, is useful in vat a and kapha, asthma, bronchitis, abdominal complaints, fevers, leucoderma, urinary discharges, tumours, piles, diseases of the spleen, pains, inflammations, leprosy, insomnia, jaundice, hiccoughs, tubercular glands and reduces biliousness (in the Ayurvedic system of medicine).

The roots and fruits are used in palsy, gout and lumbago. The root has a bitter, hot and sharp taste. It is used as a carminative, a tonic to the liver, stomachic, emmenagogue, abortifacient, aphrodisiac, haematinic, diuretic, digestive, and as a general tonic, useful in inflammation of the liver, pains in the joints, lumbago, snakebite, scorpion-sting and night-blindness (in the Unani system of medicine). In the Travancore region, an infusion of the root is prescribed after parturition, to help in the explusion of the placenta. It appears to partake, in a minor degree, of the stimulant properties of the fruit.

As an alternative tonic, long pepper is recommended for use in a rather peculiar manner. An infusion of three long peppers is to be taken with honey on the first day, then for ten successive days. The dose is to be increased by 3 peppers every day, so that on the tenth day the patient will take thirty at one dose. Then the dose is to be gradually reduced by 3 daily, and finally the medicine is to be stopped. Thus administered, it is said to act as a valuable alternative tonic in paraplegia, chronic cough, enlargements of the spleen and other abdominal viscera. Long pepper is used in the composition of several snuffs; boiled with ginger, mustard oil, butter-milk and curds it forms a liniment, used in cases of paralysis. In the Konkan region, the roasted fruits are beaten up with honey and given to treat rheumatism; they are also given powdered with black pepper and rock salt (2 parts of long pepper and 1 of salt) in half tola doses.

The dried immature fruit and the root in the form of decoction are extensively used in acute and chronic bronchitis attended with cough, and found to give gradual relief in all such cases.

Origin and Distribution

Long pepper is a native of the Indo-Malaya region. It is found growing wild in the tropical rainforests of India, Nepal, Indonesia, Malaysia, Sri Lanka, Timor and the Philippines. Indian long pepper is mostly derived from the wild plants, but is also grown in small area in the Khasi hills, the lower hills of West Bengal, Eastern Uttar Pradesh, Madhya Pradesh, Maharastra, Kerala, Karnataka and Tamil Nadu. It occurs wild in the forests of Andhra Pradesh and in the Andaman and Nicobar Islands as well.

Description of the Plant

Piper longum Linn. (2n = 24, 48, 52, 96) is a slender, climbing, undershrub, creeping and rooting below. The young shoots are downy; the leaves are 5–9 cm long, 5 cm wide, ovate, cordate with broad rounded lobes at the base, sub-acute, entire, glabrous. The plant bears unisexual flowers in solitary, erect spikes during or just after the rainy season. The male spikes are larger, slender and are 2.5 to 7.5 cm long, while the female spikes are 1.25 to about 2 cm long when in flower, growing to about 3 cm when in fruit. The berries are ovoid, yellowish orange, sunk in thick rachis about 0.25 cm in diameter.

Wild Species and Varieties

The Indian long pepper appears to be derived from two or three species, including the one which is imported from Malaysia and Indonesia. It is a product of either P longum or P pepuloides, while the java long pepper is from P officinarum DC. (Syn. Chavica officinarum Miguel). The long pepper of English commerce is chiefly the produce of the java variety. In different agroclimatic regions of the country, a number of geographical races exist. The Assam, West Bengal and Nepal races are the most popular among them. The West Bengal type is the Gol Thippali (round ball type). The Maharastra type is Pipal Nonsori.

The races found in Assam are Asali (true) and Suvali (female in Assamese). Under the All India Co-ordinated Research project on Medicinal and Aromatic Plants, a selection from 'Cheemathippalli', which bears female type spikes possessing the highest spike weight and total yield has been released under the name 'Viswam' from the Kerala Agriculture University for cultivation as a intercrop in irrigated coconut plantations.

Soil

It flourishes well in a rich, well-drained, loamy soil. It is also cultivated on a large scale in limestone soil below the Cherrapunji region. Laterite soils, rich in organic matter content with good moistureholding, capacity are also suitable.

Climate

The plant requires a hot, moist climate and an elevation between 100 and 1000 m (330–3330 ft) for its cultivation. At still higher elevations, the yield will be low. It can be successfully grown even in areas which receive heavy rainfall with high relative humidity. In its natural habitat, the plant is found growing as an undershrub. Hence, it should be grown under partial shade with 20–25% shade intensity.

Land Preparation

The land should be ploughed 2-3 times and levelled properly. Then, raised beds of size 3 m × 2.5 m are prepared, in which pits are dug at a spacing of 60 cm × 60 cm. The pits are filled with soil mixed with 100 g/pit of well-decomposed FYM or compost. In order to avoid water stagnation in the pits and beds, channels are made to drain excess rain water.

Land Preparation

Propagation

Long pepper is propagated through seeds, suckers or cuttings or by layering of mature branches at the beginning of the rainy season. The establishment percentage of 3–5 noded rooted cuttings is reported to be very high. The suckers establish very well in heavy rainfall areas. Vine cuttings can be rooted in polythene bags filled with common pot mixture.

The nursery can be raised during March and April. In the nursery, mealy bugs often attack the roots of the rooted cuttings. This can be avoided by mixing 10% DP with the potting mixture. Similarly, excessive irrigation may lead to Phytophthora, wilt of rooted cuttings. Irrigation should be done on alternate days to avoid dampness and also to control the wilt disease. The cuttings planted in March-April will be ready for planting in the main field by the end of May.

Planting

The rooted cuttings or suckers are planted in the pits at the rate of 2/pit with the onset of the monsoon. The pits are gap-filled one month after planting.

One year after planting, each plant is reported to bear 7 primary and 6 secondary roots with lengths of 12 and 7 cm, respectively. Long pepper is planted as an intercrop with subabul, eucalyptus and coconut in different parts of the country.

Manuring

This crop needs heavy manuring. Nutritional studies at the University of Agricultural Sciences, Bangalore, have shown that the application of 20 t FYM along with 50 kg N, 20 kg P_2O_5 and 70 kg K_2O/ha is required for optimum growth and yield of the crop. Of this. 50% N and the entire dose P_2O_5 and K_2O is given as a basal does and the remaining N is given as a top-dressing.

Irrigation

The crop should be irrigated once a week if it is grown as a pure crop. In case the crop is grown as an intercrop with other crops, the irrigation provided to the main crop is sufficient. The sprinkler system may be adopted for economizing irrigation water. In Kerala, it is the unirrigated crop which has shown more vigour and hardiness than the irrigated crop. It is observed that summer irrigation will induce continuous spike formation during the off-season.

Interculture

In the first year of planting, weeding is done when weed growth is noticed in the beds. After the application of FYM to the beds, earthing-up is done from the channels. Once the crop grows and Covers the interspaces of the bed, it is difficult to do any interculture operations.

Mulching

When the crop is not irrigated during summer, the beds may be mulched with dry leaves or straw.

Pests and Diseases

Pests

Mealy bugs infest the roots of plants and suck its sap. The infested plants look stunted in growth. The severity of this insect is greater in summer. The insect can be controlled by the application of a systemic insecticide like Rogor. Helopeltis theiuora is another insect whose adults and nymphs feed on tender foliage causing necrotic leisions around the feeding punctures and leaving shot-holes on the lamina.

Application of neem kernel extract at 0.25% concentration will control this insect.

Diseases

During the monsoon season, the fungus Collitotrichum spp. causes leaf and vine rotting. This fungus and Cercospora spp. cause necrotic spots and blight on leaves during the warmer months. Sometimes, these two organisms may cause partial or total crop loss. A spray of Bordeaux mixture at 1 % concentration during May and 2 or 3 sprays subsequently, during the rainy season, is the suggested control measure. Besides this, a virus-like disease leading to yellowing and crinkling of the leaves with stunted growth of plants is also reported. The affected plants produce spikes of reduced size and inferior quality.

Harvesting, Drying and Yield

The spikes are picked from the vines six months after planting. The spikes will be ready for harvest 2 months after their formation on the plants. The spikes should be picked when they are blackish-green and most pungent. If the spikes are not picked at this stage, there is a greater loss of pungency. In Bengal, picking is done annually in January, while in Kerala the spikes are picked 3 to 4 times a year. The harvested spikes are dried in the sun for 4 to 5 days until they are perfectly dry. The green to dry spike yield is around 400 kg/ha. The yield increases thereafter up to 3 years and it will be about 1000 kg/ha during the third year. After three years, the productivity of the vines decrease and they should be replanted.

Besides the spike, the thick parts of the stem and roots which have medicinal value may also be harvested from 18 months after planting. While harvesting, the stems are cut close to the ground, the roots dug up, cleaned and heaped in the shade for a day, after which they are cut into 2.5 to 5 cm long pieces. The average yield of roots is 500 kg/ha.

Grading

The dried thicker parts of the stem and roots are called piplamool. There are three grades of piplamool. Grade I with thick roots and underground stem, it fetches a higher price than Grade II and III which consists of either roots, stems or fragments.

Neem

PLANT PROFILE

Family	:	Meliaceae
English name	:	Neem tree, Margosa tree, Indian Lilac
Indian name	:	Nimbah, Prabhadrah (Sanskrit) Nim, Nimb (Hindi), Bevu (Kannada) Vembu, Vepam (Tamil)

Species and varieties	:	Azadirachta indica A. Juss
Distribution	:	India, Africa, Central and South America Andamans, Australia, Bangladesh, Carribean Fiji, Indonesia, Haiti, Malaysia, Mauritius Myanmar, Niger, Pakistan, Puerto Rico Sri Lanka, the Middle East, Sudan, Thailand Virgin Islands
Uses	:	Drugs, Fertilizers, Insecticides

Neem, Azadirachta indica A. Juss (Syn: Melia indica Brand and M. azadirachta Linn.), is one of the most valuable and yet the least exploited of tropical trees, belonging to the family Meliaceae. It is commonly known as neem tree, margosa tree, Indian lilac, white cedar, paradise tree, chinaberry or crackjack in various places.

The neem tree has great potential for agriculture, industrial and commercial exploitation because of its multiple uses such as firewood, timber and in pharmaceutical and entomological preparations. Every part of the tree from its roots, trunk, bark, leaves, flowers, fruits, seeds, sap and gum are known to have some use and have a place in the traditional folklore and medicine (especially in the Ayurvedic and Unani systems of medicine), so much so it almost plays the role of a village dispensary in this land. The leaves are reported to contain minerals like calcium, magnesium, iron and phosphorus; protein; vitamin A; crude fibre; carbohydrates; fat and alkaloids like quercetin, nimbosterol, azadirachtin, nimboflavone and nomicinol. Medicinally, its leaves are used to cure many diseases of the bladder, kidney, eyes, and skin, for the explusion of intestinal worms, and with honey for curing jaundice, boils and urticaria.

The flowers are reported to contain nimbosterol, kaempferol, quercetin, myrcetin, nimbosterin, nimocetin, fatty acids and a highly pungent essential oil. The dry flowers are considered as a stimulant, tonic and stomachic. The flowers are useful in some cases of atonic dyspepsia and general debility. The bitter oil extracted from the flowers is useful in skin diseases. The trunk bark contains a resin, nimbin, nimbinin, nimbidin, essential oil, tannins, the bitter principle margosine, desa-acetyl nimbin, inorganic phosphorus, sulphur and iron. The bark is a bitter tonic, astringent and antiperiodic. It is usually prescribed in the form of a tincture or an infusion. The bark preparations are useful in fever, thirst, nausea, vomiting, skin diseases, scorpion and snakebites. The bark is also used as a vermifuge, the aqueous extract is reported to cause immobilization of human and bovine spermatozoa. The oil of the bark possesses a marked antibacterial spectrum against Gram –ve and Gram +ve bacteria. The fresh, tender twigs are used to clean the teeth, particularly in cases of pyorrhoea.

The heartwood of the tree is slightly aromatic and is reported to yield nimatone, tannins, inorganic calcium, potassium and iron salt. Besides, two esters of sitosterol and 24-methylene cycloartanol have been found in the petroleum extract.

The pericarp of the fruit is rich is carbohydrates and constitutes a promising substrate in methane gas generation. The seeds contain calcium, magnesium, phosphorus, an amorphous morgosopicrin, diacetylnimbin, thionemone, azadirachtin, salanin, nimbin- and, vilasimin, azadiracthtol, nimbine, 6-deacetyl nimbinine, nimbanidol, nomolicinol, salannotide, meliacin-4-epinimbin, nimlin and desacetyl nimbin. The processed seed-meal contains all the essential amino acids. The seed-kernel contains about 45% of fixed oil called neem oil. The oil is non-drying, bitter and greenish-yellow to brown in colour and has a strong, characteristic, garlic-like odour.

The bitter principles of the oil are nimbidin, nimbin and nimbidol. The other active principles of the oil are salannin, meliantriol, azadirachtin, 7 -acetylneotri-chilenone, azadiradione, azadirone, 6-desacetylembinene, 17-epiazadiradione, epoxyazadiradione, framinellone, genunin, meldenin, melianone, nimbanidol, nimbidinin, nimbinene, nimbinin nimbolide, nimbolin, salannol, vepinin and volasinin. The oil contains 34 lipid associates and 48 fatty acids also. The oil displays a stimulative, antiseptic, an alternative effect when used for massage of the body and is said to possess anti-inflammatory properties. At the same time, it is an active remedy against boils and ulcers. The oil has proved to be remarkably effective in infections like ringworms, itches, erysipalis, urticaria and scrofula. In veterinary surgery, the oil is used for foul sores and it has shown antidiabetic activity in rats. Medicinally, the oil is employed as an antihelminthic, useful in bleeding gums, pyorrhoea and asthma. It is used externally as a stimulant in convulsions and collapse from fevers, cholera and fits. The oil finds use in the Siddha system of medicine also. It has antifertility and antibacterial effects. Several pharmaceutical preparations like liquors, emulsions, ointments, liniments, as well as medicinal cosmetics like creams, lotions, soaps shampoos, hair-tonics, toothpastes and gargles have been prepared with nimbidin and with the salts of the acid derived from it.

Of the salts, sodium nimbitinate has so far been used for injection purposes. Nimbidin is reported to be efficacious in the treatment of skin diseases such as warts, extensive and persistent dandruff, furunculosis, arsenical dermatitis, scabies and saborrhoeic dermatitis. Nimbidin is also known to exhibit pronounced anticholinergic, diuretic, antihistamine, antinicotinic, anti-ulcer and moderately sedative effect in experimental animals. The oil is used in manufacturing a good number of insecticides as well.

Neem-cake is obtained after expressing the oil from the seeds. The cake represents 80% by weight, of the whole seed. The cake contains oil (91.8%), nitrogen (6%), magnesium (0.88–1.3%), calcium (0.471.02%), sulphur (1.6%), proteins (37.5%), carbohydrates (25.6%), fibre (5.8%) and ash (5.4%). Besides, it also contains 110 ppm boron and tannic acid. The cake finds extensive use in crop production as an organic manure.

The gum exudes from the bark in the form of clear, bright, amber-coloured tears or fragments. It is found mixed with gumghatti. The gum contains water (13–15%), ash (3%), galactose (12%), pentosans (26%), some albumins and oxidase. The gum is a non-amphotric, polyelectrolyte and behaves like a lyophilic colloid. On hydrolysis, the gum yields arabinose, fructose, D-galactose and D-glucoronic acid. The gum also contains glucosamine and 19.8% proteins. The proteinaceous component of the gum contains 18 amino acids. The gum is prized medicinally as a stimulant, demulscent and tonic. It is useful in catarrhal and other affections which are accompanied by great weakness.

A sweet liquid, neem toddy, is occasionally obtained as an exudation from the upper parts of the trees or by exposing a healthy-looking root, cutting through it and placing a vessel beneath to receive the exuding liquor. It has the unpleasant odour of neem and contains sugar (6.5%) and albuminous and gummy matter (6.5%). The neem sap (toddy) exuded by the roots has been reported to contain glucose, sucrose, gums and protein. Its ash contains iron, aluminium and calcium. The toddy or fermented sap of the tree is useful as a refrigerant, nutrient and alternative tonic. It is of great service in some chronic and long-standing cases of leprosy and other skin diseases, tuberclosis, atonic dyspepsia and general debility. The presence of amino acids, particularly tryptophan, in fairly high concentrations is important in relation to the therapeutic value attributed to the exudate.

Various neem products, either in combination with each other or in combination with other plant products, are also used to cure many diseases.

Origin and Distribution

Neem is believed to be a native of upper Burma and possibly of the Siwalik, Deccan and other parts of South India. It is widely distributed in tropical Africa, Australia, the Andamans, Pakistan, Sri Lanka, Myanmar, Malaysia, Indonesia, Bangladesh, Thailand, Middle East, Sudan, Niger, Fiji, Mauritius, Central and South America, the Carribean, Puerto Rico, the Virgin Islands, Haiti and other places. The largest known plantation of nearly 50,000 trees is at the Arafat Plains, en route to Mecca in Saudi Arabia, for providing shade to the Haj pilgrims.

In India, it is cultivated all over. It grows wild in the Siwalik Hills, in the dry forests of Andhra Pradesh, Tamil Nadu and Karnataka. The tree is well suited for afforestation. According to one report in 1991, India has approximately 14 million neem trees and has an annual potential to produce 0.1 million tonnes of neem oil and 0.15 million tonnes of neem-cake from 0.45 million metric tonnes of seeds.

Description and the Plant

Neem is a large glabrous tree, 7 to 20 m high, with a straight trunk. The branches spread widely and form a oval crown. The trunk attains a girth of 1.8 to 2.7 m. The bark of the trunk is 1.2 cm thick, furrowed longitudinally and obliquely, dark grey outside and reddish-brown inside.

The sapwood is yellowish-white, the heartwood is red or brown, especially the inner part. From incisions in the trunk near the base, made in spring, issues a quantity of sap, often flowing for weeks. The leaves are alternate, imparipinnate, 22 to 38 cm long, the leaflets are 9 to 13 in number, nearly opposite, shortly petiolate, 2.5 to 7.5 cm long. The leaflets have unequal sides, are ovatelanceolate, sometime falcate, deep and sharply serrated, accuminate and glabrous.

The inflorescence is an axillary panicle. The flowers are white, with a strong smell of honey, especially at night, pentamerous, on short slender, pedicles with short, scattered hairs. The bracts are minute and deciduous. The calyx is small, flat, 5-fid, connate at the base, the lobes are orbicular.

The corolla has five petals, which are free, spatulate, narrow at the base and deflexed, and up to 1.5 cm long, the stamens are 10 in number, situated at the base of a hypogynous disc. The filaments form a long cylindrical tube. The ovary is 3-celled, with 2 ovules in each cell; the style is slender; with 3–5 stigma. The fruit is a drupe, ovoid-oblong; 1.25 to 1.75 cm long, smooth, greenishyellow when ripe, one or sometimes two-celled. The seed is ex-albuminous, elliptical and reticulate outside.

The germination is epigeal. The chromosome numbers as determined by metaphrase plates of meiosis of the pollen; the mother-cells are 14 in number. They are the haploid or n-chromosomes. The flowers appear from March-May and the fruit ripen during June-August.

Soil

Neem grows even in a nutrient deficient soil. It grows well on moist, dry, stony, clayey or shallow soils, including moderately saline and alkaline soils, compact clays and lateritic crusts. However, silty flats, clayey depressions and land prone to inundation are not conducive to its growth. It can tolerate a soil pH of up to 10.

Climate

The tree is very hardy and can survive high temperatures and altitudes between 50 and 100 m and little rainfall, as low as 130 mm per year, and long stretches of drought. In India, it thrives best in parts where normal rainfall ranges between 480 to 1000 mm and maximum temperatures as high as 48°C.

Cultivation

Propagation

Seed propagation is the common method of growing neem trees. It can also be propagated easily by vegetative means, like root- and shootcuttings. Root and shoot-cuttings can be made from 12–13 monthold nursery plants. A success rate of 53% from root-shoot-cuttings is reported from Maharastra.

Nursery Raising

The seeds are collected from June-August, which coincides with the rainy season. The seeds remain viable for a short period of 3 to 5 weeks only. Hence, they should be sown immediately or within 2–3 weeks of collection. It is reported that the seeds obtained 10–12 weeks after flowering possess the maximum germination capacity and longevity. The seeds can remain viable up to 2 months when stored in airtight tins, in pots with wood-ash and in gunny bags. The seeds are depulped before sowing.

The nursery is raised in beds of rows 15 cm apart with the seeds 2.5 cm apart in a row. The beds are sparingly watered after sowing. The seedlings are transplanted from when they are about 2 months till l-year-old. It is preferable to plant the seedlings when they are 7–10 cm in height. The nursery is protected against frost by providing screens. The seedlings can be raised in baskets and polythene bags also.

Planting

Direct sowing is easy and can be successfully done by dibbling, broadcast, sowing in lines or in trenches, depending on the site conditions. It is reported that, in Rajasthan, neem seed has been successfully dibbled under Euphorbia bushes in Ajmer district by placing 3–5 seeds in small pits and covering them with soil. In Maharashtra, neem is sown with 'babul'. While in the Bellary district of Karnataka, sowing on mounds in rows 2.7 m apart on black soil has proved successful.

For planting the nursery-raised seedlings, the seedlings mav be planted inside thorny bushes to provide initial protection from cattle damage. Neem can also be planted as a roadside avenue tree.

Stumps from 2-year-old plants have been reported to give a better survival and faster growth rate than those from 1-year-old plants. The stumps are prepared with 2.5 cm of shoot and 23 cm of root pieces and should be planted at the onset of the rains.

Neem in Agroforestry Systems

Neem is well suited for arid zones in India. It is the least affected by automobile exhaust and is best suited for roadside planting. It is also well-suited for afforestation of wastelands and for planting under agroforestry systems.

Neem is interplanted with groundnut, bean and millets in farmlands. It is reported that neem trees can be planted in highly alkaline soils if the land is levelled, bunded and gypsum applied @ 10-12 t/ha over the whole surface in June-July and a rice-wheat rotation followed for the next 4–5 years. The trees can then be grown along with agricultural crops.

Pests and Diseases

Pests

In India, 38 insect species are reported on neem which include seed and flower insects (5), defoliators (13), root feeders (2), stem- and shoot-borer (1), sap-sucking insects (15) and mites (2), Apart from these insects, an unidentified species of mollusca also causes 15–20% damage to the seedlings in nurseries. Parasa hilaris has been reported to severely attack neem in Madurai. The mealy bug, Pseudococus gi/beriensis, the scale insect, Parlatoria orientalis, and the leafwebber, Loboschiza koenigiana are reported to cause maximum damage in Tamil Nadu.

Among the other pests, the seed- and flower-insect Scirtothrips dorsalis prevents fruit development. The defoliating insects Boarmia variegata and Eurema spp. cause serious damage in nurseries in the arid and semi-arid regions, respectively. The main sap-sucking insects of neem are Helopeltes antonii (tea mosquito) and Pulvinaria maxima. H. antonii is reported to cause 15% damage to neem plantations in Gujarat, it is also a key pest of cashew plantations.

Holotrichia consanguinea and termites, *viz*., Odontotermes obseus, are the prominent root-feeders of the neem tree. To control the sap-suckers, including thrips, a systemic insecticide such as Dimethoate (0.01–0.2%) may be used. The larvae of the shoot-borer, L. koenigiana, can be controlled by 0.01–0.02% of Monocrotophos or Dimethoate. A mixture of 0.2% Dimethoate and Vipul is also recommended. Against defoliators, 0.1–0.2% Folithion; and for termite control, mixing Aldrex in the soil before planting, are recommended. The silvicultural measures and use of enemy complexes are also suggested.

Diseases

The plant is not attacked by any disease of a serious nature.

Harvesting and Yield

The trees shed their leaves during February-March and a full-grown tree produces an estimated 350 kg of leaves. The fruits mature in June-July and produce about 50 kg/tree of berries. The fruit contains, skin: 23.8%, pulp: 47.5%; shell: 18.6% and kernel: 10.1 %. The fruits are dried in the shade and stored in airtight gunny bags or tin containers. The fresh fruit gives, on an average, 60% on a dry-weight basis. It must be ensured that the seeds do not contain moisture, which may otherwise invite fungal attack and lead to aflatoxin contamination. The kernel yields 45% of fixed oil.

Opium Poppy

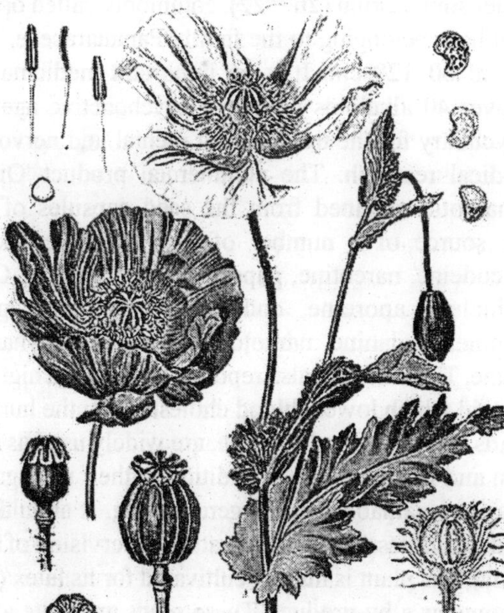

PLANT PROFILE

Family	:	Papavaraceae
EngLish name	:	Opium poppy, White poppy
Indian name	:	Ahiphenam, Aphukam (Sanskrit), Kas-kas Aphim (Hindi), Gasagase (Kannada) Kasakasa, Posttakkai (Tamil)
Species and varieties	:	Papaver somniferum

Distribution	:	India, Argentina, Bulgaria, China CzechosLovakia, Egypt, Germany Greece, HolLand, Hungary, Iran Japan, PoLand, Portugal, Spain Turkey, USSR, YugosLavia
Uses	:	Confectionery, CuLinary items, Drugs, Oil

Papauer somniferum (2n = 22), commonly called opium poppy, is an annual herb belonging to the family Papauaraceae. It grows up to a height of 60–120 cm. It is an important medicinal plant, the source of over 40 alkaloids including psychoactive agents, a great boon to psychiatry for the treatment of mental and nervous diseases and to medical research. The commercial product 'Opium' is an addictive narcotic obtained from the seed capsules of the opium poppy, the source of a number of very valuable alkaloids like morphine, codeine, narcotine, papaverine and thebain. Other minor alkaloids include aporeine, codamine, cryptopine, guoscopine, hydrocotarnine, laudanine, narcotoline, neopine, oxynarcotine and papaveramine. The seeds are also reported to contain a high percentage of linoleic acid which lowers blood cholesterol in the human system. The alkaloids, morphine and codeine, are widely used as sedatives to relieve pain and induce sleep, in addition to their use against cough. Opium is a very valuable but dangerous drug. It should be used in very limited quantities and under the strict supervision of a physician.

In India, this plant is mainly cultivated for its latex (opium) and the seeds come as a by-product. These seeds are quite a rich source of fatty oil and protein and, in many countries of Europe, employed as a major source of cooking oil. The seed is also an important culinary item in India. It is extensively used in the preparation of native confectionery, pastries and bread. In some places, the young plants are also consumed as a leafy vegetable. This is one medicinal plants which is very lucrative. A crop raised properly, in one hectare area, fetches about 1 lakh rupees. But its cultivation has to be done under the strict control of the Central Excise Department and it cannot be cultivated everywhere. It can be grown only in those areas specified by the Government of India.

Origin and Distribution

It is supposed to have originated in the western Mediterranean region and from there it has spread through the Balkan peninsula to Asia Minor and India. Since antiquity, its cultivation has been in vogue in Italy, Greece and Asia Minor. It was during the 15th century that the herb was introduced in India. First, it was cultivated along the sea coast and later penetrated into the interior of the peninsula.

Area and Production

The 1953, the United Nation's Opium Conference Protocol (still in effect) for limiting and regulating the cultivation of opium poppy plants asserts that Bulgaria, Greece, India, Iran, Turkey, the USSR, Egypt, Czechoslovakia, Poland, Germany, Holland, China, Japan, Argentina, Spain, Hungary, Portugal and Yugoslavia are the countries that may legally produce opium. Other countries might grow it for other purposes. The only country where substantial amounts of opium is now produced for export is India which amounts to 1 465 t annually and forms over 90% of the world production. In India, all the opium of commerce is now grown mainly in the states of Uttar Pradesh, Madhya Pradesh and Rajasthan covering an area of 18000 ha.

The control and regulation of all aspects of the cultivation of poppy and wholesale trade has been under the purview of the Government of the India since 1773. After the partitioning of the Indian subcontinent in 1947, the essentials of the present system of narcotics control, which derived from the system instituted by the British in the 1857 Opium Act, were retained with few changes.

The Government of India is able to effectively licence farmers to grow opium. This is highly organised so as to allow elimination of those growers whose yield of opium is poor.

Description of the Plant

The poppy is a small, erect, scarcely branched (towards the top) herb attaining a height of 120 cms with large, serrated leaves and attractive brightly coloured (white, pink, purple, red and varigated) flowers. The main shoot and branches terminate into large, oblong to globose capsules, filled with small white, flat seeds. The seeds are small, kidney-shaped, whitish-yellow, grey-brown, reddish-brown or black, and rich in oil.

Varieties

The most popular varieties grown in the country remain in the field from 140 to 160 days. The following are some of the important varieties of which Talia, Ranghatak and Dhola Chota Gothia are popular varieties recommended for heavy black soils.

Talia

It is sown early and it remains in the field for 140 days. Its flowers are pink and have large petals. The capsule is oblong, ovate, light-green and shiny (waxy).

Ranghatak

It is a medium-tall variety, maturing for lancing in 125–130 days after sowing. It bears white and light-pink flowers. It produces medium-sized capsules (7.6 cm × 5.0 cm), which are slightly flattened on the top. It yields opium of a comparatively thin consistency that changes to a dark-brown colour on exposure.

Dhola Chota Gotia

It is a dwarf cultivar (85–90 cm), bearing pure white flowers and lightgreen capsules which are oblong-ovate in shape. It is ready for lancing after 105–115 days of sowing and matures for seed in 140 days.

MOP-3

This variety has been developed at the Jawaharlal Nehru Krishi Vishwa Vidyalaya, Mandsaur, recently. It bears pinkish-white flowers comprising of large non-serrated petals. Its capsules are ready for lancing 120 days after sowing and the variety is recommended where adequate irrigation facilities exist in the later part of the season.

MOP-16

This is another promising selection made at the Jawaharalal Nehru Krishi Vishwa Vidyalay, Mandsaur. The plant bears white flowers with serrated petals and round, flat-topped capsules. This is comparatively drought tolerant and is ready for lancing 105–110 days after sowing. It is recommended where an early maturing crop is preferred.

Shama

This variety was released by the CIMAP, Lucknow during the year 1983. The main alkaloids like Morphine (14.51–16.75%), Codeine (2.05–3/24%), Thebaine (1.84–2.16%), Papaverine (0.82%) and Narcotine (5.89–6.32%) in this variety are reported to be on higher side than the existing commercially cultivated variety. It yields 39.5 kg of latex and 8.8 kg/ha of seeds.

Shweta

This variety was also released by the CIMAP, Lucknow, along with Shama. However, it is reported to be superior to Shama in the content of the main alkaloids-morphine (15.75–22.38%), codeine (2.15–2.76%), thebaine (2.04–2.5%), papaverine (0.94–1.1%) and Narcotine (5.94–6.5%). It gives an average yield of 42.5 kg of latex and 7.8 kg/ha of seeds.

BROP 1 (Botanical Research Opium Poppy-l) (NBRI–3)

It is a synthetic variety developed at the National Botanical Research Institute, Lucknow, by crossing selections from Kali Dandi, Suyapankhi and Safaid Dandi. This variety is highly adaptable to varied agroclimatic conditions and gives a higher yield than national checks. It is moderately resistant to diseases. It yields about 54 kg/ha of opium and 10-13 q/ha of seeds. The morphine content is 13% and above.

Kirtiman (NOP-4)

It was developed at the Narendra Dev University of Agriculture and Technology, Kumarganj, Faizabad, through selection from local races. The variety is moderately resistant to downy mildew. It yields 35–45 kg/ha of latex and 9–10 q/ha of seeds. The morphine content is up to 12%.

Chetak (U.O. 285)

This variety was developed at the Rajasthan Agriculture University, Udaipur. It is moderately resistant to diseases. The opium yield is up to 54 kg/ha and the seed-yield is 10–12 q/ha and contains upto 12% morphine.

Trishna (IC-42)

This variety was developed at the National Bureau of Plant Genetic Resources, Delhi, through selection and selfing. It yields 49 to 53 kg/ha of latex and 12 to 14.78% morphine. The variety is resistant to frost, root-rot and tolerant to downy mildew disease.

Jawahar Aphim 16 (JA-16)

It is a pureline selection of local land races developed at the Jawaharalal Nehru Krishi Vishwa Vidyalay College of Agriculture, Mandsaur (Madhya Pradesh). It is moderately resistant to downy mildew. It gives 45–54 kg/ha of latex, 8–10 q/ha of seeds and contains up to 12% morphine.

Recently, another three varieties:'NBRI-3' of opium, 'Sujatha' an opium-free poppy for the production of oil and seed and 'Shubhra' for high morphine and seed yield have been released from the NBRI, Lucknow, RRL, Jammu and CIMAP, Lucknow, respectively.

Soil

The opium crop needs deep clay loam, highly fertile and well-drained soils with a pH range of 6.0 to 7.5. Such soils, containing adequate organic matter, retain moisture and there is no need of irrigation during lancing.

However, with adequate manuring and use of fertilizers even light, loam to sandy-loam lateritic soils can give high yields under good management. Heavy clay or fine sandy soils generally pose a problem, since they remain wet during the rains and are too difficult to cultivate in the dry period.

Climate

In general, the crop needs long cold season (20°C) with adequate sunshine in the early season for a healthy vegetative growth; heavy rains after sowing cause loss in seed germination. Warm, dry weather with a temperature of 30-35°C is required during the reproductive period. Cloudy weather, frost, hailstorms and high gusty winds, particularly during lancing, causes immense damage to the growing crop. Dry, warm weather conditions in February-March favour a good flow of latex and results in higher yields.

Cultivation

Land Preparation and Sowing

Poppy seeds should be sown in very well-prepared soil. The field should be given 5-6 cross-ploughings, followed by planking. The land should be divided into small plots to facilitate irrigation. The seed rate is around 8 kg/ha when sown by broadcasting but when sown in lines at 30 cm apart, about 5-6 kg/ha of seeds are sufficient. In addition, this method is reported to result in higher yields, besides promoting inter-culture and lancing operations.

Experiments conducted at Mandsaur showed that a plant population of 300,000/ha is conducive to higher yields. Treating 1 kg of seed with 4–5 g of Thiram or Brassicol before sowing, protects the seeds from soil-borne diseases. The seeds are sown between late October and mid-November in different parts of the country. After sowing, the seeds are covered by a thin layer of soil followed by a light irrigation.

Manures and Fertilizers

Usually farmers maintain a rich top layer of humus by adding large amount of FYM to the opium field. It is found that the application of 10–20t of FYM is enough to maintain the soil in good physical condition and to support good initial growth. In addition, the field is applied with 30-50 kg P_2O_5 and, wherever potash is deficient, K_2O should be applied at the rate of 25–30 kg/ha. Both these fertilizers are placed 5-6 cm deep ill the rows before sowing. The amount of nitrogen varies from 60 (10 t of FYM) to 90 kg (20 t of FYM/ha). It is applied in three equal split doses, the first dose is given at the time of Sowing and the subsequent doses at 30–40 and 60–65 days after sowing. For correcting zinc deficiency, 12–30 kg/ha of zinc sulphate should be added.

Irrigation

The first irrigation is given immediately after sowing, if there is not enough moisture available in the soil. For subsequent irrigations, 7–10 days irrigation schedule is the optimum depending upon the weather and soil conditions. A total of 10 to 15 irrigations are required for this crop.

Interculture and Weeding

All cultivation operations are done by hand. When the plants are 5 to 6 cm high with 3–4 leaves, i.e., after 3–4 weeks of sowing, the first weeding is done. The intervals of the other weeding has to be adjusted locally. In all, abut 4–5 weedings may be required during the whole cropping season.

Flowering and Fruit-set

After about 90 to 100 days of sowing, the plants which are waist-high begin to flower, i.e., flowering will take place during first week of March, if the crop was sown during the second fortnight of November. Usually after 3 days of flowering, the petals fall off and after another 10–14 days the capsules are ready for lancing.

Pest and Disease

Insects

1. Cutworms (Agrotis suffusa) do great harm to these plants. They remain burrowed in the soil during day and during the nights they cut and eat the leaves. This insect can be controlled by flooding the field with water and dusting the crop with 2% Carbaryl.
2. Weevils (Stenocarus fuliginosus-root weevil and Ceutorhynchus maculalba-capsule weevil) attack young seedlings and feed on young leaves. As the plant matures, the insect eats its way up the stem and ultimately bores its way into poppy head and proceeds to destroy the capsule.

A capsule borer (Heliothis armigera) is also reported to damage the capsules.

Diseases

1. The downy mildew (Peranospora arborescens) disease may appear any time from the seedling to maturity stage. In primary infections, the plants develop chlorosis, curling, abnormal plant form with stunted growth or the plant may not even survive. If the plants are attacked in the later stage, the capsules are affected. This disease can be controlled by the application of Dithane 2–78 (0.4%).

2. Powdery mildew (Erysiphe polygoni) appears after 90–100 days of sowing as a white mycelial growth on the stem and lower side of the leaves, it grows into large patches of white powdery mass which turn black. The diseased plants in the early stage should be uprooted and burnt. At the later stage, the crop should be sprayed with Sulfex or Microsulf (1 l in 100 l of water). A spray of Dithane 2–78 or Dithane M-45 (2 g/l), beginning at the flowering stage and repeating it 2-3 times at an interval of 15 days also prevents the spread of the disease.

The other diseases reported on this crop include, root-rot caused by Fusarium semitectum, leaf-blight caused by Helminthosporium papaveris and virus diseases (cabbage-ring spot virus, beet-yellow virus and bean-yellow mosaic virus).

Lancing and Latex Collection

The lancing operation is performed by skilled labour (an average of 6 persons to a plot), usually on bright sunny days between noon and 4 p.m. The hottest part of the day is chosen, since the pellicle is said to form on the surface of the freshly exuded latex due to the hot sun, resulting in a greater degree of evaporation and quicker thickening. This also helps prevent the latex from falling off the capsule. The lancing operation is started at the edge of the field and the person works backward to avoid contact with the exuding latex. The hand is quickly passed over a capsule and a subjective decision is made as to whether it is ready for incision or not.

The lancing instrument, called "Nastar" or "Naka", comprises of four lines about the dimension of ordinary needles spaced at 1.5 to 2 mm apart and affixed to a holder about 18 cm long. The nastar is held carefully, as one holds a pencil while writing and the incision is made by a swift downward stroke starting just below the stigmatic rays. The depth of the incision is controlled by the affixation of the lines to the holder, for if incisions are too deep the latex is exuded to the interior of the capsule and is thus lost. If the cut is too shallow the yield of latex will be low. Usually, an incision with a depth of 0.4 cm is considered ideal. About 150–200 capsules can be lanced per hour by an experienced worker. Immediately on lancing, the latex exudes, it is initially milky and gets accumulated in the outer wall of the capsule. It quickly darkens and dries during the course of the day and is generally collected the next day before 10 a.m. by scraping with a trowel called the Seetoah. The collection may also be delayed for one or two days, depending upon the 'appearance' of the capsules and the vigour after the latex flow. In other words the interval is subjectively determined by the collector.

The lancing process is usually repeated twice, making a total of three series of vertical incisions. The spacing of the incisions on the capsule is generally even. If the capsules are exceptionally, large, four to five lancings can be done. The air-dried latex, which has now become blackish in colour, is scraped from the capsules into small earthenware pots which are lined with polythene sheets. Scraping is carried out by grasping the capsule between the thumb and forefinger of the left hand and inclining it gently, the scraper is then drawn upwards. The capsule is finally 'cleaned' with the thumb.

The semi-dry, blackish latex is then transferred to wooden trays and dried further. Upon arrival at the factory, appropriate samples are drawn from each grower's produce and their quality is determined in the laboratory of the Chief Opium Chemist. The material is graded according to morphine content as follows:

A = with morphine content more than 12%,
D_1 = with morphine between 11 and 12%,
B_2 = with morphine between 10 and 11 %,
B_3 = with morphine between 8 and 10%.

Processing

The crude opium is transferred to the appropriate storage vats capable of holding 3 to 30 t. The opium of variable consistency contained in these tubs is removed as required and is carefully air-dried in the sun to 70° consistency. The drying generally takes 21 days and is carried out in wooden trays, by turning the latex 3–4 times a day. Sun-drying is considered a critical process, since artificial drying causes a substantial loss in alkaloid content. Each processed tray comprises of 36 kg of opium with 70° consistency.

The opium is then packed in polythene bags; each containing 5 kg of the product, which are then shipped in wooden chests: 60 kg chest for air-freight and 50 kg/chest for surface shipment. In addition to export, a small amount of 90° consistency opium is prepared for domestic quasi-medical purposes. This is called 'Government Excise Opium' and is available in cubical cakes. These are wrapped in 'butter paper' tied and stamped and shipped to various states where the opium is taxed and sold to those who are duly registered as addicts and consumers.

Harvesting of Seeds

The capsules, after the lancing operation and collection of opium latex, are allowed to dry on the plant itself. The drying process takes about 15 days after the lancing is completed. In India, the capsules are plucked by hand and the seeds are separated after breaking the capsules. A dry capsule weigh about 7 g and it contains 11 to 12 thousand seeds weighing about 3.5 to 4 g.

Yield of Crude Opium and Seed

On an average 25–30 kg/ha of crude opium and 400–500 kg/ha of seeds are obtained in India.

Chemical Evaluation

Extraction of Alkaloids from the Poppy Husk

For this purpose, 28 ground poppy husks are treated with 20% NaCl solution (10 ml) and stirred for one hour. Then, this is further treated with a solvent mixture of methanol and benzene (1:1) (25 ml). The mixture is shaken for half-an-hour. Later on, it is filtered under suction. The residue is washed three times with a solvent mixture, taking 5 ml each time. The filtrate, thus obtained, is treated with 20 ml of HCl (0.5 N) and is kept in a separating funnel. The pH of the aqueous layer is maintained between 4–6. Now, 5 ml of HCl (1 N) is added and it is made up to 100 ml. This solution is tested by TLC for the presence of major opium alkaloids and used for the estimation of morphine.

Determination of Morphine in Poppy Capsules

Grind the dry poppy capsule to fine powder with the help of a pestle and mortar. Weigh 5 g of this dry powder into a 100 ml beaker and treat it with 25 ml 10% WN aqueous sodium carbonate solution and keep it for 1 hour. The stirred mixture is then treated with 5 ml of

20% WN aqeuous sodium carbonate solution and shaken for 1 hour with a solvent mixture (45 ml) of 3:1 volumes of chloroform and isopropanol. This proportion of the organic solvent minimizes emulsion formation. The resulting mass is suction-filtered with a 150 ml sintered Buchner Funnel and washed with 5 ml of solvent mixture, four times. The combined filtrate and washings are extracted with 1N HCl (20 ml) and then with distilled water (2 quantities each of 10 ml). The aqueous layer is filtered, adjusting the pH to 5 by treating with 1N HCl (45 ml) and diluted to 200 ml.

Take 20 ml aliquot into a volumetric flask and add 5 ml 1N HCl and make the volume 100 ml.

Take three 10 ml aliquots into three 25 ml volumetric flasks. To one add 5 ml of 0.1 N HCl, and to the other two add 2 ml Iodic acid, each. After exactly 2 minutes add 10 ml ammonium bicarbonate nickel chloride complex and make up the volume to 25 ml with distilled water. After 90 minutes, the Optical Density (red filter) in a balanced cell colorimeter-102 is measured and the morphine content computed from the calibration curve.

Extraction Procedure for Opium by Colorimetric Analysis

The sample is dried at 65° to 70°C for 24 hours and ground to a fine powder with the help of a pestle and mortar. Of this, 0.5 g (based on availability) of sample is weighed and transferred to a 15 ml centrifuge tube, containing 5 ml of water. The sample is warmed and stirred with a glass rod and kept overnight at 45°C, so that a total dissolution of all the suspension takes place and it becomes a homogeneous mixture.

To the mixture, 400 mg of calcium hydroxide is added and mixed well till a homogeneous paste is formed. It is kept for half-an-hour with intermittent stirring. Further, it is centrifuged for 10 minutes and the supernatant liquid is taken in a 100 ml volumetric flask. It is added with 10 ml distilled water to the residue in the volumetric flask, stirred well again with a glass rod and centrifuged for 10 minutes; the same procedure is repeated 6 times which gives 60 ml of combined filtrate.

It is mixed with 3 to 4 ml of 1N HCl to make the pH 4 to 5 and distilled water is added to make the volume 100 ml. 2 ml aliquot (in case of 1 g opium sample) and 4 ml aliquot (incase of 5 gm opium sample) is taken in the 25 ml volumetric flask containing 5 ml 1N HCl (pH 1.6) and the volume is made up to 10 ml with water.

Further, colour development is done similarly by adding Iodic acid and Ammonium bicarbonate nickle chloride complex and the reading is taken at 670 nm in the visible spectrophotometer after 90 minutes.

Periwinkle

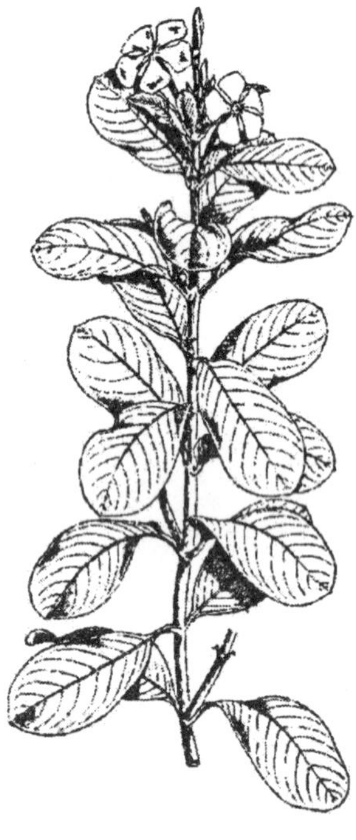

PLANT PROFILE

Family	:	Apocynaceae
English name	:	Periwinkle
Indian name	:	Madanah (Sanskrit), Mainphal, Madan Sadabahar (Hindi), Kasikanagile Smashana Mallige, Smashana Kanagale (Kannada), Madkarai, Marukkalankay (Tamil)

Species and varieties	:	Catharanthus roseus (L) G.Don Nirmal, Dhawal
Distribution	:	India, Indo-China, Indonesia, Israel, Madagascar, Philippines, South Africa, USA
Uses	:	Cosmetics, Drugs

Periwinkle (Chatharanthus roseus (L.) G. Don), belonging to the family Apocynaceae, is one of the few medicinal plants which has found mention in the folk medicinal literature as early as 2nd Be. The plant has been widely used as an abortificent, purgative, antidiabetic, diuretic, haemorrhagic, antimalarial, antidysentric and against skin diseases by the ancient people.

Modern investigations have shown that periwinkle contains more than 100 alkaloids, distributed through all the parts of the plant. It has medicinal importance owing to the presence of alkaloids like ajmalicine (raubasin), serpentine and reserpine which are well known for their hypotensive and antispasmodic properties. Periwinkle gained further importance after the isolation of vincristine and vinblastine alkaloids which have importance in cancer therapy. Vincristine sulphate is being marked under the trade name ONCOVIN, which is used against acute leukemia, and vinblastine sulphate as VELBE to cure Hodgkin's disease and other lymphomas and choriocarcinomas. In addition to the above, the alkaloids leurosidine, leurosovine and rovidine also possess anticancer properties, but they are not used clinically.

Origin and Distribution

The plant is a native of Madagascar and from there, it has spread to India, Indonesia, Indo-China, Philippines, South Africa, Israel, USA and other parts of the world. In India, it is being grown in Tamil Nadu, Karnataka, Andhra Pradesh, Madhya Pradesh, Gujarat and Assam in an area of about 3000 ha. Farmers prefer it because of its wide adaptability and its ability to grow on marginal lands and drought-tolerance that rules out crop failure. The presence of alkaloids all over the plant body confers immunity to cattle-browsing and the crop loss due to pilferage. USA is the world's largest user of this plant's raw material. A single firm which has the patent to manufacture Vinblastine and Vincristine sulphate have been consuming more than 1000 t of leaves of the plant annually.

Most of it has been imported from Malagasy and the remaining from India and Mozambique. Hungary is also been one of the major consumer of its leaves. West Germany, Italy, Netherlands and the UK are interested in the roots. They extract ajmalicine (raubasin), serpentine, etc. The total demand from these countries is more than 1000 t of roots annually.

Description of the Plant

C. roseus (2n = 16) is a perennial herb which grows up to 90 cm high. The stem is erect, lax branching with flexible long branches, purple or light-green in colour. The leaf is simple, opposite, exstipulate, petiolate. The inflorescence is racemose, with the flowers in axillary pairs. The flowers are rose-purple or white or white with a rose-purple spot in the centre, pedicillate, hermaphrodite, complete, actinomorphic and hypogynous.

There are five calyx-sepals, five corolla-petals, rose-purple or white or white with a rose-purple spot in the centre, tubular, swollen in the region of the anthers, the throat of the corolla-tube is hairy, forming a corona-like structure. The androecium has five stamens included in the corolla tube, the filaments are very small, epipetalous, the anthers forming a cone-like structure above the stigma. The gynoecium has two carpels, free below but united in the stylar region, two ovaries, free, single style with a dum bel-shaped stigma. After fertilization, the carpels separate and form two fruits, which form a pair of elongated follicles. The seeds are small, black in colour, 1 g contains about 835 seeds.

Types and Varieties

There are three variants in Periwinkle, those with (i) rose-purple flowers, (ii) white flowers, and (iii) white flowers with a rose-purple spot in the centre. The first type is being cultivated because of its higher alkaloid content. Recently, two white-flowered varieties named 'Nirmal' and 'Dhawal' have been released by the CIMAP, Lucknow, which, although, equal in active principles are reported to yield a higher biomass.

Soil

The crop is quite hardy and grows well op a wide variety of soils, except those which are alkaline or water-logged. Deep sandy loam to loam soils of medium fertility are preferred for its large-scale cultivation. Because, in this soil there is not only a better development of roots, but it is also easy to take them out at harvest time.

Climate

The distribution of the plant shows that there is no specificity in its climatic requirements. It comes up well in tropical and subtropical areas. However, the growth in tropical areas is better than in the subtropical areas, where its growth is slow due to the low temperatures In winter. It can be successfully grown up to an elevation of 1 300 m above sea-level. A well distributed rainfall of 100 cm or more is ideal for raising this crop on a commercial scale under rainfed conditions.

Propagation

The plants can either be propagated by seeds or vegetatively through cuttings. Since plants propagated by cuttings flower earlier than the plants from seeds, it is recommended that for drug production the plants should be grown from seeds and for seed production from cuttings.

Propagation by Seeds

Fresh seeds collected a few months in advance are preferred for sowing, as they lose viability on long storage. In an experiment, it was observed that the seeds had an initial viability of about 95 to 96% but when stored under uncontrolled conditions for two years, there was no germination. The seeds can either be directly sown in the field, or a nursery can be raised and the seedlings transplanted.

Direct Sowing

This method is best suited for large areas where labour is expensive as it reduces the cost of production. For this purpose, the land is ploughed twice and brought to a fine tilth. Weeds, stubble and pebbles are removed. The field is divided into plots of convenient size and the soil is mixed with the recommended does of manures and fertilizers. The seeds at the rate of 2.5 kg/ha are broadcasted at the onset of the monsoon in June-July, in lines spaced 30–45 cm apart and lightly covered. Since the seeds are very small, for ease in handling and distribution, they are mixed with sand about 10 times their weight. Germination takes place after about 7–8 days. After the germination is complete, the seedlings are thinned at a spacing of 30–40 cm within the row. The plant may grow upto a height of 120 cm and the flowering starts 40–45 days after sowing.

Nursery Raising and Transplanting

When seed supply is short, this method can be followed. The other advantage of this method, in comparison to direct sowing, is that we can select the healthy and vigorously growing seedlings for our purpose and the inferior ones can be discarded. The seeds are sown in well prepared, raised nursery beds in March-April by broadcasting them in rows spaced at 8–10 cm apart and about 1.5 cm deep. They are then covered with a light soil and leaf-mould mixture and watered to keep the bed moist. About 500 g of seeds will be enough to raise seedlings to cover a one hectare area. In about 2 months time after germination, the seedlings are ready for transplanting into the field. The seedlings are transplanted at a spacing of 45 cm × 30 cm apart in the field. With this spacing, one hectare requires a population density of about 74000 plants.

Vegetative Propagation

To raise the plants by this method, soft-wood cuttings obtained from the lateral roots have proved better than either the hard or semi-hard wood cuttings. Cuttings of about 10–15 cm length with a minimum of 5–6 nodes, have been reported as being ideal and result in about 90% rooting. Soaking the cuttings over-night in NAA solution of 25 or 50 ppm concentration has been found to further improved rooting to the extent of 96%. This method can be profitably used for multiplying the clones which have high alkaloid content and also where we have to produce only the seeds.

Manures and Fertilizers

In areas where FYM is available, it is applied at the rate of 10–15 t/ha to obtain good growth and yield from periwinkle plants. If irrigation is available, it is recommended to grow leguminous crops like sun hemp or horsegram and when they reach the flowering stage bury them inside the soil before sowing or transplanting periwinkle. Green manure will act as a substitute for FYM and is useful in the areas where it is either difficult to procure or it is very expensive. The seeds of the green manure crop should preferably be treated with bacterial innoculants prior to sowing, to increase the development of root nodules which absorb atmospheric nitrogen and fix it in the soil. In case organic manure is not applied, it is advisable to apply a basal dose of 20 kg N, 30 kg P_2O_5 and 30 kg K_2O per hectare per year. In addition, a top-dressing with 20 kg nitrogen can be given in two equal split doses during the season.

Irrigation

In places where rainfall is evenly distributed throughout the year, the plants do not require any irrigation. However, in areas where rainfall is restricted to a few months in a particular period, about 4–5 irrigations will help the plants to give optimum yield.

Weed Control

This crop requires two weedings in the initial stages of its growth. The first weeding may be done after about 60 days of sowing and the second at 120 days of sowing. Mulching the field with cut-grass or rice-straw will also minimize the weed growth.

Application of the chemical weedicide Sinbar at 4–5 kg/ha as a preemergent spray is highly effective against all monocot weeds.

Similarly, the application of a mixture of 2–4–0 and Gramaxone at the rate of 25 kg/ha to the soil before sowing keeps the weeds under control.

Insects, Pests and Diseases

The plant is sufficiently hardy and practically free from the attack of insect pests and diseases. However, the oleander hawk month is reported on this crop. Occasionally, some plants have been found to suffer from the 'little leaf' disease, due to infection by mycoplasma resulting in stunted growth and rosetting of the leaves of the plant. The disease can be effectively checked by uprooting and destroying the affected plants.

Recently, another disease 'dieback' or twig blight or top rot caused by Pythium butleri, Phytophthora nicotianae, P. debaryanum,

Alternaria tenuissima and Colletotrichum dematium has been found to affect the crop during the monsoon in some parts of the country. The disease can be controlled by spraying Oithane 2–78 at an interval of 10–15 days.

The other fungal diseases reported on this crop are Fusarium wilt caused by Fusarium solani and Sclerotium ro lis ii, blight caused by Pythium aphanidermatum as well as leaf-spot caused by Myrothesium roridum, A. tenuissima, A. alternata, Rhizoctonia solani, Ophiobolus catharanthicola, Haplosporella marathwadenis and Glomerella cingulata.

Harvesting

Roots

The crop is harvested after 12 months of sowing. The plants are cut about 7.5 cm above the ground level and dried for the stems, leaves and seeds. The field is then copiously irrigated and when it reaches the proper condition for digging, it is ploughed and the roots are collected. The roots are later washed well and dried in the shade.

Leaves, Stem and Seeds

If there is a demand for leaves, two leaf strippings the first after 6 months and the second after 9 months of sowing can be taken. A third leaf stripping is also obtained when whole plant is harvested. After the plant is harvested, it is dried in the shade. Its seed-pods dehisce and release the seeds with a light threshing, which can be used for the next sowing.

The leaves and stems are also collected separately. It may be mentioned here, that the seeds collected in this way will have poor germination because they have been collected from pods in different degrees of maturity. Therefore, in order to obtain good seeds, it is advisable to collect them from mature pods two to three months before the harvest of the crop. The aerial part of the plant between 7.5 cm and about 25 cm above the ground level, is taken as the stem for the purpose of marketing.

The total alkaloid content in the leaf varies from 0.15 to 1.34%, of which the average content of vinblastine is 0.002%, while that of vincristine is 0.005%.

Yield

Under irrigated conditions, about 4 t/ha of leaves, 1.5 t/ha of stem and 1.5 t/ha of roots, on an air-dried basis may be obtained. Whereas, the yield under rainfed condition may be about 2 t/ha of leaves and 0.75 t/ha each of stem and roots on an air-dried basis.

Identification and Estimation of Vincaleukoblastine

Vincaleukoblastine (VLB) is a potent antineoplastic drug, present in the leaves. It is a dimeric alkaloid containing indole and incoline moieties. For the estimation of VLB, the sample of leaves is dried at 60°C to constant weight and powdered in a dry grinder. Out of this, about 5 g leaf-powder is extracted with 50 ml methanol in a metabolic shaker at room temperature for one hour and left overnight. The extract is filtered and the residue is re-extracted three times with 50 ml methanol each time for half-an-hour. The combined methanol extract is concentrated to 20 ml in a vacuum, diluted with 20 ml water, acidified with 20 ml sulfuric acid and extracted twice with 20 ml ethyl acetate saturated with water.

The ethyl acetate fraction is rejected. The aqueous phase is made alkaline with 10% ammonia and the liberated alkaloids are extracted with dichloromethane (3 × 50 ml). The combined dichloromethane extracts are dried over anhydrous sodium sulphate, filtered and evaporated to dryness in a vacuum. The residue is dissolved in 20ml benzene and extracted with 50 ml 0.1M aqueous citric acid. The pH of the acidic solution is adjusted to 4.4 with 10% ammonia and the liberated alkaloids are extracted thrice with 50 ml of dichloromethane each time. The combined dichloromethane extract is dried over anhydrous sodium sulphate, filtered and concentrated to dryness in a vacuum.

The residue containing the fraction of alkaloids liberated at pH 4.4 s dissolved in 2 ml of chloroform and 0.2 ml of it is applied as a 3-4 band on precoated silica gel Silgur-25 UV_{254} TLC plates. The pates are developed in ethyl acetate-absolute ethanol (1:1). After development, the plates are air-dried in the dark and, when viewed under long wave ultraviolet light, a dull yellow fluorescent band at R1 0.33 is observed which is identical with the R1 reported for VLB earlier, under similar conditions of TLC.

The band acquires the characteristic orange colour of alkaloids on spraying with Dragendorff's reagent. Corresponding bands from other plates are scraped along with blank silica gel portions of same area and eluted three times with a total of 10 ml ethanol. The measurement of optical density of the combined elutes at different wavelengths in a double-beam spectrophotometer show that the compound has maximum absorption at 214 nm. This, again, is a property which is characteristic of VLB. From the standard equivalence of optical density and the VLB concentration reported in literature, the VLB content of the leaves could be worked out as 6.8 mg/100 g on a dry-weight basis. The results are the average of estimations on a triplicate set of samples.

The 5 to 20 III of alkaloids extracted at pH 4.4, dissolved in acetonitrile and 0.01 M diammonium hydrogen orthophosphate, which is also the mobile phase, is analysed by HPLC. The alkaloid fraction, thus analysed, reveals the presence of a compound with the same retention volume reported for VLB under similar conditions of HPLC.

Podophyllum

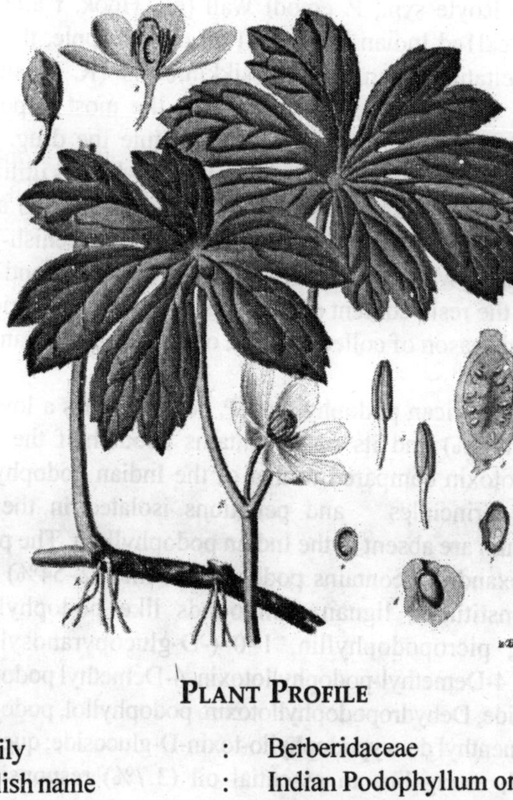

Plant Profile

Family	:	Berberidaceae
English name	:	Indian Podophyllum or Mayapple
Indian name	:	Bakra-chimyaka, Papra, Papri (Hindi)
Species and varieties	:	Podophyllum Linn. P. hexandrum Royle (Syn. P. emodi Wall) P. peltatum Linn.

		P. sikkimensis P. axillans, P. bhootanensis P. hexandrum, P. jaeschkei
Distribution	:	India, Afghanistan, eastern Canada Pakistan, USA
Uses	:	Drugs

The genus Podophyllum Linn., belonging to the family Berberidaceae, has three economically important drug plants P hexandrum Royle syn., P. emodi Wall (ex. Hook. f and Thomas), commonly called Indian Podophyllum or Mayapple, the American plant P. peltatum Linn. and P. sikkimensis (R. Chatterjee and Mukerjee), of which the first species is the most important. The rhizomes and roots of these plants constitute the drug. The dried rhizomes and roots contain a medicinal resin which is utilized almost entirely for podophyllin preparation. Podophyllin is an amorphous power varying in colour from light-brown to greenish-yellow or brownish-gray, with a characteristic odour and bitter and acid taste. Generally, the resin content of the roots and rhizomes varies with the locality and season of collection and, on an average, it ranges from 7 to 15%.

The American podophyllum (P. peltatum) has a lower content of resin (1–8%) and its resin contains about half the amount of podophyllotoxin compared to that of the Indian podophyllum; also the active principles and peltations isolated in the American podophyllum are absent in the Indian podophyllum. The podophyllin from P hexandrum contains podophyllotoxin (32–54%) as a major active constituent; lignan compounds like podophyllotoxin-D-glucoside, picropodophyllin, 1–0–(-D-glucopyranosyl) picropodophyllin, 4-Demethyl-podophyllotoxin, 4-Demethyl podophyliotoxin -D-glucoside, Dehydropodophyllotoxin, podophyllol, podophyllic acid and 4-Dementhyl deoxypodophyllo-toxin-D-glucoside; quercetin (8%), kampferol, astragalin, an essential oil (3.7%) responsible for the odour of podophyllin, wax (8.5%) and mineral salts.

Podophyllin is considered a cholagogue, purgative alterative, emetic, vermifuge and bitter tonic. It can be used like colchicine in cytological work, in the form of ointment in the treatment of venereal and other warts, in veterinary medicine, in the treatment of certain types of cancers, *etc*. Recently, the resin from the American podophyllum has been found to affect the growth of malignant tumors and it has also proved effective in controlling some skin diseases.

Origin and Distribution

P. hexandrum is indigenous to the Himalayan regions of India, Pakistain and Afghanistan; and P. peltatum to shady and moist places in eastern Canada and USA. P hexandrum grows wild in the interior ranges of the Himalayan region of India at altitudes of 2700 to 4200 m from Sikkim Hozaro descending to 1828 m (6000 ft) In Kashmir, Simla, the higher ranges of Kangra, Kulu and Chamba. The rhizomes and roots of this plant are collected from natural sources in the forests. The experimental cultivation of this plant was started long time ago at the Minor Forest Products Garden at Chakrata, Dehra Dun, and by the Cinchona Department of Tamil Nadu. Owing to several problems in its cultivation, not much work has been done on this crop and the information available in literature on its cultivation aspects is meagre.

Description of the Plant

Podophyllum is an erect, herbaceous plant about 35 to 60 cm in height with a creeping perennial root system and annual aerial parts, which appear after the middle of March after its dormancy during winter. The leaves are alternate, long, petiolated, often purple-spotted, around 15 to 25 cm in diameter, deeply divided in the middle or base into 3 to 5 lobes which are sharp-toothed and often have deep incisions.

The peduncle is terminal in the bud stage, apparently supraxillary or inserted on the petiole of the upper leaf. The flowers are 3.8 to 5 cm in diameter and white to purple in colour. The fruit is a large, scarlet, pulpy berry, 2.5 to 6.3 cm in length, containing 40 to 80 seeds embedded in the pulp. P. hexandrum (2n = 12) exhibits a certain amount of variation in its botanical features, particularly in the nature of leaf insertion. They are classified into three or four varieties, *viz.* var. hexandrum, var. axillaris, var. bhootanensis and var. jaeschkei.

Description of the Economic Parts

The rhizome is irregular, tortuous, knotty, about 2–5 cm long and 1–2 cm thick, there are 3 or 4 cup-shaped scars on the upper surface; externally, the colour is yellowish-brown to earthy brown. The transversely cut surface appears smooth and is irregularly circular in outline.

Its odour is slight and characteristic; its taste is rather bitter and acrid. The powder is intensely irritating to the eyes. The roots are adventitious and arise mainly from the enlarged portion of the rhizome and attain a length of about 7 cm and a thickness of 2.5 mm. The fracture of both the rhizome and root is weak, brittle and even. The odour of the root is faint and its taste is disagreeably bitter and acrid.

Soil

It flourishes well in soils rich in humus and decayed organic matter in its natural habitat. The plant prefers medium or light-medium loam, fertile, well-drained soils for its cultivation.

Climate

In the fir forests of the Himalayas, Podophyllum is found growing as undergrowth. The plant is generally associated with species of Rhododendron, Saliz, Juniperus and Viburnum and less frequently found in open alpine meadows.

The plant prefers temperate, moist and shady localities situated between 2500 and 4000 m.

Seasons

The rhizome cuttings may be planted from May to the beginning of July. However, planting at the end of June or the beginning of July gives the best results. The seeds may be sown in June or July.

Land Preparation

The land may be ploughed well two to three times until a fine tilth of soil is obtained. Before the second or third ploughing, FYM @ 8-10 t/ha may be applied.

Cultivation

Propagation

Podophyllum is propagated by seeds and also vegetatviely from rhizome cuttings. For propagation, rhizome cuttings are preferred as seed germination is poor.

Propagation by Seeds

The seeds have a dormancy period of about one to two years. They may even fail to germinate if they are sown without the pulp. Pre-sowing treatments like soaking the seeds in water or treating them with sulphuric acid have no favourable effect on germination. The seeds may be sown immediately after their collection. They germinate in 7–10 months and the percentage of germination will be from 42 to 44.5%. The seedlings are transplanted into the field when they are 6–10 cm tall. The crop raised from seedlings usually takes a number of years to produce fair-sized, marketable rhizomes.

Propagation by Rhizome Cuttings

Rhizome cuttings, 1 to 2.5 cm in length, taken from the youngest top portions of the rhizome, bearing leafy buds, gives the best results both in the percentage of sprouting and growth of sprouts. However, the growth of plants propagated by rhizome cuttings is also slow and the plants take 4 to 6 years to produce rhizomes suitable for exploitation.

Planting

When seedlings or rhizome cuttings are used for planting, they may be planted at a spacing of 15–20 cm × 30–35 cm or 30 cm × 30 cm. If the crop is to be raised by the direct sowing of seeds, freshly extracted seeds at the rate of not less than three seeds should be dibbled at the same spacing to ensure uniform stock. The seedlings may be thinned out when they are 5–7 cm high, retaining only one plant per hillock.

Manuring

It is desirable to give a dressing of well-rotted and sieved farmyard manure by the end of October or the beginning of

November. A fertilizer mixture consisting of 20 kg N, 40 kg P and 40 kg K/ha should be applied at the time of planting. In addition, 60 kg/ha N should be applied in two split doses as a top-dressing during the growth period.

Irrigation

The crop may be irrigated during the dry period, depending on the soil type and weather conditions. Generally, 2–4 irrigations are required.

Weeding and Hoeing

Weeding and hoeing may be done as and when necessary. Generally, one or two weedings are required to be done during the rains. Hoeing may be done along with the second weeding.

Pests and Diseases

No serious incidence of any insect pests or diseases are noticed on this crop. However, rhizome-rot disease is sometimes noticed, for which suitable control measures may be taken up, by treating the rhizomes with a suitable fungicide.

Harvesting and Yield

The development of rhizomes in this plant is very slow. The rhizomes are ready to be collected for the market only when they are 2 to 4 years old. The yield of rhizomes and roots is reported to be high if the crop is harvested 5 or 6 years after planting. In this plant, the underground rhizomes remain dormant during winter and produce aerial shoots in April or May. The shoots bear flowers and fruits during summer and die down in November. Usually, the rhizomes which bear 3-5 aerial shoots are considered suitable for harvesting. In spring or autumn, such plants are dug up and the rhizomes and roots are separated and cleaned. Then they are cut into longitudinal pieces and carefully dried. The rhizomes harvested in spring are reported to contain a higher resin content than those obtained in autumn, and also the freshly-harvested rhizomes contain large quantities of active principles which are lost on prolonged storage. The yield of dried rhizomes and roots varies from 2.5 to 5 t/ha.

Pritnrose

PLANT PROFILE

Family	:	Onagraceae
English name	:	Evening Primrose
Indian name	:	Dheriani (Hindi)
Species and varieties	:	Oenothera lamarckiana L.
		O. biennis
Distribution	:	India, temperate America
Uses	:	Cosmetics, Drugs, Health food, Oil

Primrose (Oenothera lamarckiana L.), commonly called the evening primrose, belonss to the family Onagraceae. This is a beautiful biennial plant of hybrid origin that has produced many other lovely varieties, and can best be described as a dwarfish form of Oenothera biennis only with much large flowers and bigger leaves, which are wide and crinkled. Its stems are reddish and the flower-buds hairy. The name Oenothera is derived from the Greek word, oinos (wine) and thera (hunt or chase). Plants of this genera are native to North and South America.

Evening primrose is grown for its seeds from which oil containing gamma linolenic acid (GLA) is extracted. The seed normally contains 20 to 25% oil, of which 8 to 10% is GLA. The GLA-rich oil is marketed principally as a health food that is used to treat various ailments such as cardiovascular disorders, metabolic disorders, dermatological conditions, diabetes, breast problems, premenstural syndromes, actopic eczema, rheumatic arthritis and alcoholism.

Distribution

The genus Oenothera is distributed chiefly in temperate America, with some species occurring in the tropics. Some species have been introduced into Indian gardens, of which few have run wild.

Description of the Plant

It is a biennial herb. Parenchyma is present in submerged stems, mostly respiratory roots are present on the upperside of the submerged stem. The leaves are simple, alternate or opposite, entire or have a toothed margin, the stipules are absent or caducous. The flowers are solitary, auxiliary or arranged in racemes, spikes or panicles. The perianth is biseriate, the floral axis is elongated beyond the ovary into a calyx tube or hypanthium. The calyx has four sepals, which are free or the hypanthium is extended above the ovary, is mostly persistent and valvate. The androecium consists of seven stamens, the anthers are two-celled, dehiscing longitudinally. The gynoecium is composed of four carpels, rarely 2–5, is syncarpous; the ovary is inferior; axile placentation, the style is long and filiform, the stigmas are usually capitate. The fruit is a capsule, and the seed is endospermic with a straight embryo.

Soil

Although the crop can grow well on a wide range of soils, a fertile loam soil is considered ideal. The established crop has an excellent tolerance to drought and is capable of growing on soils of poor fertility. However, the plant is sensitive to water-logging and good drainage is critical for its survival. If the crop is being directly seeded, soils which are prone to crusting should be avoided.

Climate

It prefers a moderately cool climate for its successful growth. A soil temperature of 20°C is ideal for the yield and the quality of the oil. Higher temperatures not only decreases the seed yield but also the content of the gamma linolenic acid in them. However, in recent years, some high-temperature tolerant types have been isolated, leading the possibility of growing this crop in comparatively warmer regions also.

Cultivation

Propagation

Primrose can be directly seeded or transplanted. Although seeding is cheaper, establishment problems and a higher potential for weed competition makes direct seeding more risky than transplanting.

Direct Seeding

To sow the crop by this method, the field is thoroughly prepared to bring the soil to a fine tilth. It is later laid out into convenient-sized blocks to facilitate irrigation. The seeds are usually sown using a precision seeder, in early August. About 1 kg of seeds are enough to cover 1 ha area. Before sowing, the seeds are mixed with 10 kg of sand for easy distribution. The seeding depth should not exceed 1.5 cm. Low moisture content in the soil can severely reduce seed germination and affect seedling emergence. Light irrigation, if available, should be given to supply the moisture necessary for germination.

The optimum density for this plant to grow and yield commercially is about 30 plants per square metre. Extremely dense population should be avoided as the crowded plants may not flower.

Wherever such a situation is present, the plants are thinned to achieve a population that will allow successful flowering. A thin population on the other hand, is of much less concern since the plant has the ability to produce many side branches.

The other method of direct sowing suggested is to dibble about 4-5 seeds at a spacing of 60 × 30 cm to thin out the seedlings retaining one plant per spot after 20–25 days of sowing. About 1 kg of seeds are required to cover 1 ha for this method of sowing also.

Transplanting

The transplanting method, although more costly, has certain advantages, like: the crop is more easily established, the desired population is more readily achieved and weed control problems are minimized, and intra-row cultivation is possible.

For transplanting, 0.5 kg seed should be sown in well-prepared, raised nursery-beds of convenient size during the month of August. When the seedlings are about 60 days old they are transplanted into the main field on ridges, in such a way that a spacing of 90–100 cm between the rows and 30 cm between plants is maintained. A spacing of 60 × 30 cm is also recommended.

Manures and Fertilizers

As this is a newly domesticated crop, very little information on the nutrient requirement of this crop is available and there is a need to work it out locally. However, in the absence of any other recommendation, FYM @ 2 t/ha and a fertilizer dose comprising of 40 kg N, 60 kg of super phosphate and 40 kg/ha of muriate of potash may be used to raise this crop.

Irrigation

It does not require much water. Initially, for 2–3 months, irrigation is provided at 4 to 5 days' intervals. After this period, when the crop has developed the proper root system, the interval is increased to once in 10 days.

Cultural Operation

Weeds are kept under control through regular weeding and hoeing, since there are no herbicides recommended for this crop.

The crop must be kept free from weeds, because weed seeds are difficult to separate from the primrose seeds and they bring down the quality of the oil, making it unacceptable to the market.

Pests

Cut-worms are reported to attack the plants during the early period of growth, damaging the foliage seriously and causing a setback.

Tarnished plant bugs and aphids are the insects which appear in late June or early July, and continue feeding throughout the remainder of the season, retarding the growth of the plant. Spraying the crop with Carbaryl @ 4 g/l of water controls these pests.

Diseases

Botrytis spp., a variety of fungus, may attack the new growth in early spring, resulting in the death of the plant. It is considered as a potential problem on crops caught with untimely heavy rains during the growth season. Drenching the soil with 0.2% of Bavistin helps to control the disease.

Harvesting and Yield

The specific growth habit of the plant indicates variance of seed maturity on each plant. The maturity of pods is observed at the bottom and the top ones are the least mature. As individual pods ripen, they begin to open and shed the seed. Thus, the timing of harvest must be carefully chosen to minimize losses and, at the same time, efforts must be made to allow as many seeds as possible to mature.

The best way to judge the stage of maturity of the crop is. to observe the opening of the pods, progressively, from the base of the stem towards the top. Physiologically mature seeds at the bottom are dark reddish-brown or black in colour with rough surfaces, as you move up the boundary between mature and immature seeds. Mature seeds may also be present in green pods, since the drying up of the pods is the last stage of the maturation process. The crop is normally considered ready for harvesting when the bottom 5 to 6 seed pods are split and start shedding seeds and the upper third of the plant is light-brown in colour.

The crop would normally be harvested in April-May. It must be harvested and allowed to dry for 5 to 7 days before threshing and winnowing. The seeds should contain not more than 10% moisture for safe storage. A well-maintained crop of primrose may yield about 800 to 1000 kg/ha of dried seeds.

Rauvolfia

PLANT PROFILE

Family	:	Apocynaceae
English name	:	Rauvolfia root, Serpentina root Serpentine root
Indian Name	:	Sarpagandha, Chandrika (Sanskrit) Chandrabhaga, Chota chand (Hindi) Sarpagandha (Kannada) Sarppaganti, Sivan amalpodi (Tamil)

Species and varieties	:	Rauvolfia serpantina Benth. ex kurz R. tetraphylla RS-1
Distribution	:	India, Andaman Islands, South-east Asia Bangladesh, Burma, Indonesia, Malaysia, Sri Lanka
Uses	:	Drugs

Rauvolfia serpantina, Benth. ex Kurz (2n = 22) (Family: Apocynaceae), commonly referred to as Sarpagandha, is one of the most important native medicinal plants of India. Apart from R. serpentina there is another species, R. tetraphylla which is also cultivated on a commercial scale. In India, the root of sarpagandha has a 400 year history of use in treatment of snakebite, insect stings, nervous disorders, mania and epilepsy; intractable skin disorders, such as psoriasis, excessive sweating and itching; gynaecological ointments for menopause, toxic goitre and in conditions such as angina pectoris and to promote uterine contraction in childbirth. The importance of the drug and the alkaloids obtained from it has been recognized by the allopathic system in the treatment of hypertension and as a sedative or tranquillizing agent.

The hypotensive properties of rauvolfja was first discovered by Chopra in 1953, but the attention of the Western countries was drawn to it only after the isolation and identification of its most active alkaloid 'Reseripine' by the Swiss scientists, Schilier and Mueller, of CIBA Pharmaceuticals in Switzerland, in 1952. After that, there was a great demand for its roots, leading to its indiscriminate uprooting from wild sources where it grew luxuriantly. Intense collection brought the plant to the verge of extension. So, in 1955, the Government of India put a ban on the export of the raw drug and attempts to cultivate the plant were taken up at a number of places. About 30 alkaloids are known to exist in this plant. The most important among these are rescinamine, doserpidine, reserpinine, serpentine, serpentinine, ajmaline, ajmalicine, ajmalinine, rauvolfinine and yohimbine. The total alkaloid content varies from 1.7 to 3% of the dried roots. Detailed studies have been carried out on the chemistry of these alkaloids, their pharmacodynamics and their varying roles in essential hypertension and neuropsychiatric conditions.

Origin and Distribution

R. serpentina is indigenous to the moist, deciduous forests of South-East Asia including Burma, Bangladesh, Sri Lanka, Malaysia, the Andaman Islands and Indonesia. Most of the drug is obtained from wild sources in these countries. It is cultivated on a small scale in India and Bangladesh.

In India, it is found in the central region, i.e. between Sirmor and the Gorakhpur district of Utter Pradesh, in shady, moist or sometimes swampy localities. In the east in Bihar, North Bengal and Assam as well as in Khasi, Jaintia and Gharo Hills the plant is encountered more commonly on the forest margin of mixed deciduous forests. In the Western Ghats, R. serpentina occurs more frequently in Goa, Coorg, the North Kanara and Shimoga districts of Karnataka and Palghat, Calicut and Trichur in Kerala. In Orissa, Andhra Pradesh and Himachal Pradesh the areas comprising the catchment of the river Godavari are the richest. The plant is chiefly associated with Sal (Shorea robusta) forests as well as bamboo brakes.

A major part of the commercial supply of this drug used in the USA and European countries originates from India, Pakistan, Sri Lanka, Myanmar (Burma) and Thailand, with India being the major supplier. The present day commercial supplies of the roots of rauvolfia are mostly from Uttar Pradesh, Bihar, Orissa, West Bengal, Assam, Andhra Pradesh, Tamil Nadu, Kerala, Karnataka and Maharastra. Rauvolfia roots exhibit more or less the same morphological features but they differ in their alkaloid content, not only in geographically distinct areas but also within the same geographical area. The annual requirement of roots in the country for the manufactured rauvolfia extracts, simple and total alkaloids, have been estimated at about 650 t against the present annual supply of 30 t from all the sources. There is also a great demand for the alkaloids as well as the raw drug in the international market. The world requirement of dried rauvolfia roots is around 20000 t/annum. The consumption of the raw drug is substantial in the indigenous drug market. Steps, therefore, will have to be taken to increase the present production to about 100–150 t/annum of dry roots. This is possible only if the plant is brought under large-scale cultivation in agro climatically suitable areas.

Description of the Plant

The plant is a perennial undershrub, growing to a height of 60-90 cm. Its leaves are simple, 7.5–10 cm long-and 3.5–5 cm broad, eliptic or lanceolate, glabrous, bright-green above and pale-green beneath, pointed and occurring in whorls of 3-5. The inflorescence is a many flowered corymb with white or pink flowers. The fruit is a drupe, 0.5 cm in diameter and shiny black when fully ripe. The root system consists of a prominent, tuberous, soft tap-root reaching a length of 30–50 cm in a 2-year-old plant. Its diameter at the thickest portion varies from 1.2 to 2.5 cm. The root-bark, which constitutes 40-60% of the whole root, is rich in alkaloids. The fresh roots emit a characteristic acrid aroma and are very bitter in taste.

Varieties

As a result of the evaluation of six populations from different locations, a culture was purfied and designated as 'RS-l' and released for commercial cultivation by the Jawaharalal Nehru Krishi Vishwa Vidyalay, College of Agriculture, Indore. The RS-1 culture gives 50-60% seed germination even after storing it for seven months and the yield of air-dried root gives upto 25 q/ha, containing 1.641 to 2.94% of total alkaloid at eighteen months of age.

Soil

The plant grows in a wide variety of soils, from sandy alluvial loan to red lateritic loan of stiff dark loan. In its natural habitat, it prefers clay to clayey loan with a large percentage of humus, and its does not grow well in soil having pH 8 or above. The ideal pH for this crop is from 4.6–6.2; generally, the plant produces thicker roots in black, stiff loan soils than in heavy clayey or sandy soil. Soils containing large quantities of sand, rated the growth of the plants and make them more susceptible to root and leaf diseases.

Climate

A climate with a temperature range of 10–30°C seems to be well suited for this plant. The best areas for its growth are those which combine high rainfall with properly drained soil. Although it has been reported to grow naturally where rainfall measures about

250 cm annually, it grows well in areas with a rainfall of even 500 cm or more. In low rainfall areas, the plant can be successfully cultivated, if irrigation is available during the drier months. Though the plant seems to be sensitive to water-logging, it can withstand water for 2 to 3 days without too much damage. The plant sheds its leaves during the cold months in localities with severe winters. Frost kills the top tender, green twigs only and fresh shoots sprout up with the advent of spring from the thicker shoots which can withstand the frost.

Cultivation

Propagation

Rauvolfia can be propagated by seeds and also by vegetative means like root-cuttings, root-stumps, stem-cuttings and leaf-cuttings.

(A) Seed Propagation

Commercially, the plant is usually propagated by seeds. Irregular and low percentage of germination of seeds is the main difficulty in the propagation of rauvolifia. The percentage of germination of seeds is quite variable, ranging from 10–60%. This is partly attributed to the adverse influence of the stony endocarp. Another serious factor is the absence of embryos, though the fruits may appear perfectly normal externally. This may possibly be an effect of parthenocarpy or of deferred somato-plastic-sterility. The rate of germination, however, depends on the percentage of fully matured, heavy seeds in a particular lot. Fresh seeds, collected from ripe fruits and immediately sown, show a higher percentage of germination (58–74%). The ripe seeds collected from the beginning of June to the end of October or even November and stored in airtight bins, retained their viability for about 6 months. The viability of the seeds drops markedly with the increase in the interval of time between collection and sowing. The germination rate of the seed also differs under varying agroclimatic conditions.

The collection of mature seeds is usually done from September-February. Fruits mature between July-November. Only a few fruits ripen at a time and, if they are not collected immediately, they shed and are lost. Therefore, the collection of ripe fruits twice a week is necessary. While this is easy in case of plantations, collection from plants growing in the wild is both laborious and costly owing to their scattered distribution over large areas. The fruits, after collection, are freed from their pulpy covering by rubbing them against old gunny bags or on rough flooring. The cleaned seeds are thoroughly dried in the sun and stored in dry places or in airtight containers. A yield of 100–120 kg of clean seeds can be obtained from one hectare of a 3-year-old plantation. An adequate supply of seeds can also be obtained by raising the plants vegetatively in compact areas, preferably by using stem-cuttings.

Direct sowing of the seeds in the field has not been successful and hence, seedlings are raised in the nursery and transplanted into the field. The nursery should ideally be located in partially shaded areas with irrigation facilities. The land is cleared of weeds and ploughed to a depth of 30 cm. Raised beds are made, which should contain one-third quantity of well-rotted FYM and two-thirds of fine soil. Seeds are sown in the middle of May. The seeds should be soaked in water overnight and the light seeds which float can be discarded. The seeds can be treated with Thiram at the rate of 3g/kg of seeds. About 5.5 kg of seeds sown in a 500 m^2 area will yield seedlings sufficient to plant one hectare. The germination is gradual and the growth of the seedlings is slow. Germination starts after 15–20 days and continues up to 40–50 days after sowing. The nursery should be kept moist throughout the germination period.

(i) Transplanting

Seedlings of 40–50 days which have 4–6 leaves are ready for transplanting. The seedlings are carefully dug out and the tap-root should be cut. They are then dipped in a 0.1 % solution of emisan fungicide before planting, to protect them against soil-borne fungus causing damping-off disease. Well-rotted FYM at 25–30 t/ha is added during land preparation. The field is then divided into small plots for irrigation. About 15 cm deep furrows are dug at a distance of 45 cm. The seedlings are tranplanted into the furrows, by making holes large enough to receive the seedlings along with the accompanying clump of earth. A spacing of 30 cm between the plants should be maintained. The seedlings are buried up to the first pair of leaves and the soil around them is lightly pressed. Irrigation after transplanting is essential, and should be continued at regular intervals until the seedlings are established.

(B) Vegetative Propagation

As collection of seeds from wild sources is both laborious and costly, vegetative propagation by root or shoot-cuttings has been advocated for raising plantations to collect the seeds, as well as to quickly multiply the genetically superior clones.

(i) By Root-cuttings

Large tap-roots with a few filiform lateral secondary rootlets are used. Cuttings of 2.5–5.0 cm length are planted in holes, at the begining of the monsoon and are almost completely covered with earth, leaving only 1 cm above the surface. Nearly 50% of the root-cuttings sprout in about a month. Trials have shown that under irrigated conditions root-cuttings of about 0.25 cm diameter planted during March-June give a 50–80% success rate. About 100 kg of root-cuttings are required to plant 1 ha. The high percentage of success obtained by using root-cuttings makes it more preferable than propagation by seeds. However, the recovery of roots from them has been found to be not as high as in plants raised from seeds.

(ii) By Root-stumps

Propagation by using about 5 cm of root with a portion of the stem above the collar has also been attempted. This method gives about 90–95% success, sometimes even 100%. Such plants transplanted in May-July into irrigated fields become well established by the end of September. This method has its limitations, as only one plant can be raised from a single stump.

(iii) By Stem-cuttings

Stem-cuttings taken from woody twigs have also been tried as a source of propagation. Hard-wood cuttings have been found better than soft-wood cuttings. Cuttings of 15–22 cm length, with three internodes are the most suitable. Nearly 60–100% of rooting is obtained by treating hard-wood cuttings with -indole acetic acid solution of 30 ppm for 12 hrs and the treated cuttings root within 15 days. Stem-cuttings planted in the nursery during the early monsoon (June) and kept moist until they sprout, give about 40–65% success rate. Such cuttings, though they start sprouting 3–4 days after planting, naturally strike roots mostly after about 75 days. Stem-cuttings have been found less statifactory than root-cuttings, since many of them do not root easily.

Manuring

The use of organic manure, leaf-mould and compost has been recommended to increase the quantity of nutrients in the soil and improve the drainage. The plants respond better to chemical fertilizers than to organic manures. Nitrogenous fertilizers induce more vegetative growth, followed by organic manure. Nitrogen seems to have a stunting effect on the root. But the combination of nitrogen either with FYM or phosphates results in better root growth than nitrogen alone. Application of phosphates induces more growth of thick as well as thin roots. Since good manure is in short supply and unecomomical, artificial fertilizers should, therefore, be preferably used. It is better to apply 25–30 t of well-rotted FYM at the time of land preparation and 10 kg N, 60 kg P_2O_5 and 30 kg K_2O/ha as a basal dose. Later, two equal doses of N, each of 10 kg/ha in moist soil is given at 50 days and 170 days after planting.

Irrigation

The crop may be irrigated fortnightly in the hot dry season and about once a month in winter. The crop can be cultivated under rainfed conditions also, but the yield is considerably poorer.

Weeding and Interculture

In order to maintain the satisfactory development of roots about 2 weedings are necessary during the monsoon and one hoeing at the end of the growing season (December). This may be done in large plantations using a tractor-drawn cultivator which is cheaper than manual labour. Hoeing by means of a tractor-drawn wheel-hoe, is the most economical.

Intercropping

It is possible to grow intercrops in rauvolfia plantations, particularly where good irrigation facilities are available. It is reported that, although the yield of roots was higher under monoculture, the net returns were highest when rauvolfia was intercropped with soyabeans and onions or soybeans and garlic.

Pests and Diseases

Insect Pests

1. Root knots appear as galls of various sizes, covering the rootsystem. Stunted growth of plants, etiolation and decrease in leaf-size are the symptoms in the aerial portion. The galls, on examination, show the presence of mites, various soil fungi and nematodes (Heterodera sp.). Generally, brown clay soils supporting the natural growth of R. serpentina are found to be infested with nematodes, while dark clay soils are free from them. Application of 25 kg of 3 G Carbofuran or 20 kg of 10 G. Phorate granules per hectare will control them.
2. A pyralid caterpillar (Glyphodes uertumnalis) causes appreciable damange to the leaves. The caterpillars roll the leaf and feed on the green matter. Some other caterpillals (Daphnis nerii, Land Deilophola nerii, L) are also reported to feed on tender leaves, causing defoliation of the plant. They can be controlled by spraying 0.2% Rogor.
3. Cockchafer grubs (Anomala polita, Blanch) attack the seedlings about 2 cm below the hypocotyl, resulting in their drying up. To control the attack of grubs, mix phorate granules with the soil at the time of nursery prepartion. Haygrub (Scarabidae) is also reported on rauvolfia.

Diseases

1. Leaf-spot casued by Cercospora rauuolfiae manifests as dark-brown coloured spots on the upper surface of the leaf and yellowish-brown on the lower Surface. The affected leaves turn yellow, become dry and subsequently falloff, resulting in defoliation. To control this disease Dithane 2-78 or Dithane M-45 @ 0.2% is to be sprayed in early June, before the monsoons and repeated at montlhly intervals until November.
2. Alternaria tenuis attacks the leaves, resulting in minute, brownish or dark-coloured circular spots with a yellowish margin on the ventral side of the leaves; these spots enlarge to form prominent dark-brown circular lesions. The fungus also affects the

flowers and fruits. The crop should be sprayed with 30 g Blitox in 10 litres of water, whenever the symptoms are seen.

3. Mosak is a common disease on R. serpentina. The primary symptom comprises of vein-banding and the gradual yellowing of the leaves, later the leaves curl from the margin to the mid-rib on the abaxial surface and, ultimately, the leaves drop off. Strict selection of seedlings at the nursery stage helps in avoiding the disease.

The other diseases reported include target leaf-spot caused by Coryneospora cassicola and Pellicularia filamentosa, leaf-blotch caused by Cercospora serpentina, anthracnose caused by Collitotrichum Cultivation and Agronomy of Medicinal Plants Colliitotrichum gloeosporoides, die-back caused by Collitotrichum dematium, powdery mildew caused by Leviellula taurica and fusarium wilt (Fusarium oxysporum f. rauvolfii). The root-knot nematode (Meloidogyne spp.) is also reported on this crop.

Harvesting and Processing

The roots of exploitable size are generally collected 2-3 years after planting, Le. from 18 months onward. It is reported that roots dug out in winter (December), when the plants have shed their leaves, are richer in total content of alkaloids than the roots harvested in August. A light irrigation should be given in advance to facilitate easy digging of roots. The roots may be dug out carefully from the subsoil, manually or by using a mould board plough. Care should be taken to keep the root-bark intact as the bark constitutes 40–56% of the whole root and has a higher alkaloid content. The roots are freed from the adhering soil, washed if necessary and thoroughly air-dried till they become brittle and are usually packed in gunny bags. Roots containing 8% moisture keep better. They are stored in a cool, dry place to prevent mould. When the roots are to be stored for a short period, they may be packed in gunny bags or polythene bags, pending disposal. But these must be periodically turned, to protect them from mould which is reported to reduce the total alkaloid content ranging from 5–25%. Under the present system, only tap-roots are selected for processing. It has been observed that the rootlets are also rich in alkaloids, so these should be included in the material.

Yield

Though R. serpentina can be propagated by various methods, the optimum yield of roots (including thick, thin and fibrous) is obtained when the propagation is done by seeds. The yield of fresh roots per plant varies widely from 0.1–4 kg. With a spacing of 60 × 30 cm, the total yield of roots in the case of plants raised from seeds works out to about 1 175 kg/ha on an air-dried basis, as compared with 175 kg/ha in the case of plants raised from stem-cuttings, and 345 kg/ha in the case of plants raised from root-cuttings. A yield of 2 200 kg/ha of air-dried roots has been obtained from a 2-year-old plantation and 3 300 kg/ha from a 3-year-old plantation, under irrigated conditions on sandy, clay loam soil.

Roselle

Plant Profile

Family	:	*Malvaceae*
English name	:	Hibiscus, Shoe-flower
Indian name	:	Lalambri, Patwa (Hindi), Pundi, Pundigida (Kannada), Pulichaikeerai, Gogu Mandharam (Tamil)
Species and varieties	:	*Hibiscus sabadrifja* L. *sabdarifja*

Distribution	:	India, Central America, California, Cuba Egypt, Florida, Indonesia, Malaysia, Nigeria Philippines, Sri Lanka, Sudan, West Indies
Uses	:	Confectionery, Drugs, Foods, Wines

Roselle (Hibiscus sabadriffa L. var. sabdariffa), belonging to the family Malvaceae, is an important multipurpose annual undershrub preferred for its red, acid, succulent calyces, young leaves, stem, fruits, seed and seed-oil for food, beverages and in Unani and modern systems of medicine throughout the tropics and subtropics, the world over. Apart from its wide range of uses, roselle is becoming increasingly important because of the growth in its export as a source of natural dye, obtained from the dried calyces. The calyces which are valued as the main items of trade contain acids, pectin, mucilage, crude protein, carbohydrates, ascorbic acid, anthocyan ins and minerals.

The fresh calyces contain 88.26% moisture, 3.74–4% acid, 3.19% 'pectin, 1.45% crude protein, 1.97% ether extracts, 5.86% carbohydrates, crude fibre, 0.87% ash, 0.108% calcium, 0.052% phosphorus, 0.021 % iron, manganese, aluminium, sodium and potassium in traces, 0.82% reducing sugars, 0.29% sucrose, mucilage, gossypetin, hibiscitin chloride, ascorbic acid and anthocyan ins. The principal water-soluble acids present are citric acid, d-malic acid; tartaric acid and hibiscus acid. Hibiscitrin chloride, on hydrolysis, yields glucose, pentose and delphinidin chloride. Cyanidin-3-glucoside, Delphinidin-pentoside-glucoside, Delphinidin-3-sambubioside and cynadin-3-sambubioside are the water-soluble pigments in the anthocyanins. Among these Delhinidin-3-sambubioside and cynadin3-sambubioside are the major pigments. The other two are present only in traces.

The seeds contain 12.9% moisture, 3.29% nitrogen, 16.8% fatty oil, 16.8% cellulose, 15.5% pentosans, 11.1% starch, myricetin, quercetin, 3.28% protein and 21.1 % lipids. In addition, eighteen amino acids have also been identified and quantified. The most essential amino acids are leucin, lysine and phenyl-alanine. Tryptophan, valine, isoleucine and threonine are the most limiting. Oleic acid is the most predominant fatty acid, followed by palmitic acid and stearic acid, while potassium, sodium, magnesium and calcium are the major elements present in the seeds. Gossypol is found only as a trace.

The dried calyces have acquired a considerable reputation in the international trade as a substitute for tea and coffee 'and as a popular soft drink. The fresh calyces are used as a vegetable or eaten as salad or used in confectioneries, in the preparation of soups, curries, chutneys, sauces, jams, jellies, squashes, tarts, tisanes and wines. The astringent herbal drink made from hibiscus, known as medicinal tea, is used as a cure for fever, cold, cough, pneumonia, and is said to possess diuretic, demulscent, emollient, laxative, antiseptic, antiscorbutic, antispasmodic and choleretic properties. The drug is also used to cure heart and nerve diseases, asthma and skin diseases. Besides, the seeds contain fatty oil, proteins, starch and lipids. The seed-meal is made into soup. Medicinally, the seeds are reported to have demulscent, diuretic and tonic properties and are recommended in cases of dyspepsia, strangury and general debility. The oil of the seed is said to have antibacterial properties.

Origin and Distribution

Tropical Africa is considered as its place of origin. It is also being cultivated in warm countries, particularly in the Philippines, Malaysia, Indonesia, India, Cuba, Central America, California, Florida, Egypt, Sudan, Nigeria, West Indies and Sri Lanka. In India, it is cultivated in Punjab, Bihar, Uttar Pradesh, West Bengal, Assam, Orissa, Maharastra, Karnataka, Andhra Pradesh and Tamil Nadu.

Description of the Plant

Roselle is an erect, branched, glabrous undershrub, 0.5-1.8 m tall, with a strong tap-root. The stem is red or green in colour. The leaves are alternate, the stipules subulate, deciduous, 5-8 mm long. The lamina is variable. The lower leaves are often ovate and undivided, others palmately 3-5 lobed or, rarely, 7 -lobed. The lobes are oblong, lanceolate, 7–15 cm long with the middle lobe being the longest, the margin is serrate with a gland near the base of the predominant mid-rib. The flowers are solitary, axillary, 3–5 cm long with a grayish-yellow corolla and an orange staminal column. The peduncle is short. The epicalyx has up to 10 linear, fleshy bracteoles. The calyx is 5-lobed, conical, 1–2 cm long, enlarging and becoming fleshly after anthesis. The epicalyx and calyx are red, whitish or green in colour. There are 5 petals, 3–5 cm long. The staminal column is 1–2 cm long with numerous anthers. The stigma is 5–lobed. The fruit is a ovoid capsule, villous, 2–3 cm long, closely invested with an enlarged, persistent fleshy calyx, dehiscing by 5 valves when ripe. The seeds are reniform, dark-brown, pilose, 4–6 mm long and are 22--34 per capsule.

The flowers open late in the morning and close at mid-day, remaining open for more than three hours. The stigmas are flush with the opening of the staminal column and do not blow away into the air. The filaments are very short and the anthers burst around the opening of the column. Self-pollination is favoured by these arrangements and the completion of the process is brought about by the closing of the flower. Flowering has a dual effect on growth. When floral induction is strong, lateral buds develop into flowers and' the number of branches will be proportionately reduced. When the first fruits are harvested early, the resting flower buds grow and the total flowering period is lengthened.

Studies have revealed that Hibiscus sabdariffa L. var. sabdariffa has a chromosome number of $2n = 4 \times = 72$ ($2n = 36$ is also recorded) and is a tetraploid.

Soil

The crop is not very exacting in its requirements and can tolerate poor conditions and is well adapted for growth in a wide range of soils. But laterite soils and light soils deficient in humus have proven to be unsuitable for the satisfactory growth of this plant. Hence, the soil selected for its cultivation should be permeable, not very compact and not subject to flooding. Sandy soils and heavy loam soils capable of retaining moisture are also suitable. However, for obtaining viable economic yields, a loamy soil with a good quantity of organic matter is ideal.

Climate

Roselle prefers a warm humid or dry climate with an even rainfall of 150 to 200 cm yearly and about 25 cm per month during the growing period. It can also grow in drier rainfed areas as an irrigated crop, but it cannot withstand heavy continuous rains, water-logging, winter cold and frost it is known to do well from locations at sea-level up to an elevation of 600–780 m. The monsoon crop is usually sown from the beginning of April to the end of May. In some places where the temperatures during winter do not go very low, it is also grown from October-November.

Roselle is a short-day plant requiring 12–12.5 hours light in the period before flowering. It is found to show ambiphotoperiodic reaction flowering both in short days and in extremely long days, remaining vegetative (at least at 21–25°C) in a 16-hour day.

Cultivation

Sowing

Roselle is usually propagated by sowing the seeds directly into the field. As the seeds lose the viability rapidly, seeds from the previous season's crop only should be used for sowing. About 2-3 kg of seed is required to cover one hectare.

Before sowing, the field is thoroughly prepared by bringing the land to a fine tilth with repeated ploughings. It is then laid out into plots of convenient sizes by constructing bunds and channels. Generally, a plot of 6 m × 4.5 m size is preferred as it helps in the proper management of irrigation, weeding and other intercultural operations. After the preparation of the plots, the soil is incorporated with 5 t of well-decomposed FYM or compost. Ridges and furrows are made at a spacing of 75 cm along the rows and the rows are irrigated lightly immediately after sowing. In some cases, thin or thick medium-sized shoot-cuttings are also used for propagating it vegetatively.

Manures and Fertilizers

Roselle responds well to the application of manures and fertilizers. Well-decomposed FYM, at the rate of 5 t/ha, is incorporated into the soil at the time of land preparation. Subsequently, a fertilizer dose of 50 kg N, 75 kg/ha each of P and K and is given at the time of sowing. An additional quantity of 200 kg N is applied in two equal split doses as top-dressing at 30 days and 70 days after sowing.

Irrigation

After· sowing, the plots are immediately provided with light irrigation. Later, irrigation is given at regular intervals till the seedlings emerge and are well established. In the crop sown in the rainy season except during some long dry spells, practically no irrigation is required. However, the crop sown in October-November needs to be irrigated once in 7–8 days, depending on the weather conditions.

Interculture

Interculture in this crop is mostly confined to weeding. Two weedings during the early period of growth are enough to suppress the weeds. As it has a good canopy in the later stages, the weeds do not get much of a chance to grow.

Pests and Diseases

Insect Pests

The crop is very hardy and there is no report of any serious insect pest in roselle. However, some insect pests like capsule borer, flea beetle and the cotton stainer bug have been observed to affect the crop. Two sprays of 0.2% Quinolphos or 0.15% Monocrotophos will control the attack of the capsule borer and flea beetle. The spraying of any contact insecticide like 0.4% Carbaryl or 0.2% Endosulphan will control the incidence of the cotton stainer bug.

Diseases

Black-foot or Stem-rot

This disease is caused by the fungus Phytophthora parasitaica var. sabdariffae, due to which the plants suddenly wilt. It is usually severe at the reproductive stage, particularly when there is a high moisture content in the soil due to heavy-rains.

Root-rot

The soil fungus Rhizoctonia bataticola, alone or with Fusarium oxysporum, causes this disease. This is also severe under wet soil conditions. Improving the soil drainage and controlling irrigation reduces the incidence of these diseases. Treating the seeds with

0.2% Captan or 0.1 % Calixin or Kitazin before sowing and then drenching the plants with 0.1% Carbendiazim or 0.2% Diafoltan or 0.3% Mancozeb chemicals a fortnight after germination, controls the disease.

A virus disease sometimes causes siight damage to the fruiting plants. The leaves of the affected plants become hard and crack when crushed. The affected plants may be uprooted and either burnt or buried to avoid further incidence of this disease.

Harvesting

Roselle is a 5–6 months duration crop. The plants come to the flowering stage after about 45–60 days of sowing and it extends over a period of 4 months. The c psules begin to mature progressively from the lower to the upper portion of the plant. The calyces grow rapidly and are ready for picking within 15–20 days of blooming. The growth of the calyx depends upon the production of phytohormones in the developing seeds.

While harvesting, the whole fruits are plucked when they are tender, plump, fleshy, crisp and deep-red in colour. The calyces are then separated by hand. If the capsules are allowed to grow older, the tissues become tough and harvesting is difficult. The calyx lobes gathered this way are placed fresh or dried in the shade for 12-15 days till a moisture level of 12% is reached. The harvesting season lasts for 2–3 months.

For seed production, the calyces are picked while the capsule is still attached to the plant. After some time, when a few of the lower and middle capsules turn dry and start splitting, exhibiting wel! developed seeds, they are harvested separately or the plants are cut at ground-level. After harvest, the plants are allowed to dry in the field for some time. They are then brought to the threshing yard and the seeds are separated. These seeds are cleaned by winnowing and packed in airtight containers till they are marketed.

Extraction of Dye and Oil

Although the extraction of dye from the calyces on a commercial scale is yet to be exploited, on a small-scale it is being done, using small samples of fresh or dried calyces. To extract the dye, the fresh calyces are ground in a wet grinder with water, boiled in an open container, strained and repeatedly washed with water and the filtrate is collected. The brilliant red filtrate is reduced to the required volume by boiling and is cooled and preserved in bottles or tins. However, for the extraction of dye from the dried calyces, overnight soaking in water is necessary and thereafter a procedure similar to the extraction of dye from fresh calyces is followed.

Organic solvents like ethanol and methanol can also be used in place of water for the extraction of the dye. The method for estimation of anthocyanin as suggested by Fuleki and Francis (1968) is the standard official method which may be followed.

Yeild

An yield of 13–14 t/ha of fresh calyces which in turn may yield 1.8-2 t/ha dried calyces, may be obtained. The recovery of the water-soluble extracts containing the dye may be around 1.97% on a fresh-weight basis.

In addition, a seed-yield of 2.5–3.0 t/ha, which contains 16.2% fatty oil, is also obtained from this crop.

Rutin-bearing Eucalyptus

Plant Profile

Family	:	*Myrtaceae*
English name	:	Red Stingy bark, Eucalyptus
Indian name	:	Nilaniryasa, Tailaparnah (Sanskrit) Yukeliptas (Hindi), Yukkalimaram (Tamil)
Species and varieties	:	*Eucalyptus macrorhyncha* E. *youmani*
Distribution	:	India, Australia, New Zealand
Uses	:	Drugs, Oils

Eucalyptus macrorhyncha, commonly called Red Stringy-bark Eucalyptus is one of the six thousand-and-odd species of the family

Myrtaceae. It is considered to be the best source of rutin, both in terms of quantity and quality. This plant being perennial, and due to the ease of cultivation even on moderate soil under rainfed conditions, offers great advantage in commercializing its cultivation in India. There are vast areas and different regions in the country which can adopt this crop to meet the rutin requirements indigenously. Furthermore, E. macrorhyncha is considered the best source of the drug, because the rutin produced from it is considerably purer and requires less capital outlay when compared with other sources. The leaf is the main source of rutin in this crop.

Rutin is being produced on large scale in Australia and New Zealand from E. macrorhycha and its sister species, E. youmani. The rutin content in the leaves of these species is of the order of 8–10%.

Rutin is an important pharmaceutical product, useful in the treatment of capillary fragility, retinitis and possibly, rheumatic fever of haemorrhagic conditions. Generally, capillary fragility occurs in a significant number of cases of high blood pressure, diabetes and other conditions. It is also useful in the prevention of apoplexy and retinal haemorrhage, in those cases where capillary fragility is concerned. It is used in cases of coronary thrombosis and for the purification of blood. Rutin also protects against the harmful effects of X-rays, indicating that it may be of use to persons exposed to dangerous atomic radiations. Under certain conditions, rutin can protect animals against histamine shock.

This pharmaceutical product is not found indigenously in sufficient quantities in our country. Presently, the rutin requirements of India is estimated at 7 t annually and is being met mainly through imports from Australia, involving huge sums of foreign exchange.

Origin

E. macrorhyncha is indigenous to the Central and Southwestern slopes of the New South Wales and Victoria provinces in Southern Australia.

Area and Productions

In Australia, E. macrorhyncha grows at 300–900 m elevation from MSL. At present, this species is commercially cultivated on vast areas for rutin production. The product is exported to several countries of the world. In our country, its trial cultivation was carried out in the Palani Hills in Tamil Nadu. The Regional Research Laboratory (RRL), Jammu, has recently taken up its cultivation at Palampur Station in Himachal Pradesh and has developed agrotechniques for its commercial cultivation. In Karnataka also, there are many potentially suitable areas for its successful cultivation on commercial scale.

Two units have been set up on Himachal Pradesh for the manufacture of rutin on a commercial scale with technical know-how from the RRL, Jammu. This new introduction in the country, if taken up for large-scale cultivation, will benefit both the growers and the country as a whole in meeting the indigenous needs of rutin as well as for export to other countries.

Description of the Plant

E. macrorhyncha is a perennial, medium-sized tree, attaining a height of 20–30 m, ½ to 1 m diameter, with a dark-reddish to brown bark, which is persistent and stringy. The leaves of the mature tree are typical of the genus and the crown is moderately less dense. The leaves are alternate, petiolate, narrow, lanceolate and somewhat venulose. The leaves have very thick cuticles and the laminae are borne vertically downwards. The leaves contain oil glands and tannin cells. The lifespan of the leaves is short in this species, compared with the evergreen trees of other genera. The mean life of the leaves is about 18 months. It varies with their position on the tree as, on vigorous branches, many leaves fall during the first year and most of the remainder fall with the new flush. Generally, the leaf-fall is a constant process and is greatest at the commencement of periods of fresh growth and when the fruit is forming.

Climate and Soil

E. macrorhyncha prefers altitudes between 300 and 900 m (under Australian conditions). Trials conducted at different elevations in the range of 500 to 1 500 m MSL, to find the optimum altitude for large-scale plantations show clearly that altitudes higher than 900 m are not favourable, as the species does not exhibit the juvenile phase perfectly. In Australia, it is grown in less fertile lands, in regions subject to hot spells and droughts. This species is reported to improve the texture and fertility of land. It is envisaged that in Karnataka, areas around Bangalore, Hassan and Chikmagalur which are less useful for other crop production can be profitably utilized for the cultivation of this species. The area selected should be comparatively cooler and elevated.

Land Preparation and Planting

The main field is prepared well by clearing the weeds and stubble. One or two ploughings are given well before planting. The square method of planting is adopted. Pits of 15 cm are dug in the field at a spacing of 1.5 × 1.5 m for the transplantation of the seedlings. On an average 4400 /ha is the ideal plant population. About 6 to 12 g of Gramaxone should be incorporated in each pit along with FYMI compost mixture (2: 1), before transplanting the seedling into the pits. The pits are well filled with the mixture at least one or two weeks before transplanting so that it settles down. Planting is done with utmost care, as the seedlings are delicate and small. The seedling bags are carefully kept in baskets and transported to the field and placed one near each pit in the main field. A hole is scooped out in the centre of the pit to the approximate volume of the bag containing the seedling. The polythene cover is carefully removed without disturbing the root-system. The seedling along with its ball of earth is planted in the hole, and the soil pressed from sides to secure the seedling well in the pit. The planting in pits is done in late June-July when the monsoon is in full swing.

Cultivation

Propagation and Raising of the Nursery

The crop is raised from seeds which are very small in size (720 seeds/g). The germination of seeds in this crop is very low even under ideal conditions, ranging only from 4 to 6%. The seeds normally lose viability on aging and, so, fresh seeds are used for raising the nursery. Nursery-raising is necessary as the tiny plants require intensive care to protect them from frost, severe cold and winter rains. Further, it takes a pretty long time, almost 6 months, for the seedlings to be fit for transplanting, when they attain a height of 12–25 cm.

The nursery is started in January-February in shallow seed-pans or trays. The container is filled with a sand and soil mixture, enriched with well-decomposed organic matter. The seeds are sown by broadcasting them in the container evenly and covering them with a thin layer of sand. The seeds sown deeper than 1 cm usually do 110t germinate. Generally, screening of healthy seeds is done before sowing. The seed-trays are well covered with paddy-straw after sowing to ensure uniform watering and heat conservation.

The seedlings obtained from sowing during January-February are quite healthy, less prone to diseases and attain a considerable height before transplantation in the mid-monsoon period. If sown during the hotter periods like April-May, they suffer from damping-off, stunted growth and chlorosis. The germination is fairly good in the June-August period, but the plant growth is stunted in the nursery stages. Hence, January-February sowing is recommended. The straw layer is removed immediately after the commencement of germination, after 20 to 25 days. The tiny seedlings having a pair cf opposite leaves, are picked out tenderly and transplanted with care. A secondary nursery of polythene bags is also maintained. The plants are ready for transplanting into the main field in about 5–6 months' time.

Sometimes the seeds are sown shallow, directly in the polythene bags, packed with well-mixed soil and sand enriched with organic matter. An average of 10–12 seeds are sown individually in each bag to ensure that at least one germinates in each bag. The bags are covered with paddy-straw until germination. In case more than one seed germinates in a bag, the extra seedling is picked out in the early stage and transplanted into a separate bag. The seedlings are maintained in the bags for 5–6 months under intensive care and then transplanted into the main field at the right time.

Manures and Fertilizers

To supplement the nutrient requirements an adequate N, P, K mixture 15:15:15 at the rate of 10–15 g per plant is incorporated in the soil around the young seedling twice during January-February and again in August-September each year during the first two years. From the third year onwards, nitrogen alone at the rate of 220 kg/halyear in the form of urea is placed around individual plants, in two split does after each harvest. Frequent foliar spray, consisting of N, P, K and minerals at the rate of 0.5% concentration increases the foliage yield considerably.

Irrigation

Protective irrigation is essential during the dry periods of October November and May-June. It also helps the plant's establishment and profuse growth. This two-fold purpose is achieved by giving one or two irrigations during each dry season.

Weeding and Interculture

Weeding and hoeing are necessary during the first year after transplanting. The area around each plant is completely dug out for better aeration to the root-zone and also to control weed growth.

These operations are carried out manually after the monsoon season is just over.

Training and Coppicing of Plants

Frequent coppicing (plucking of leaves from crown portion), is done in the second and third year, regularly. It is desirable to maintain the height of the plant at a moderate height of 2.0 m to 2.5 m only, by pruning away the terminal shoots beyond 2.0 m. These two practices help in developing a denser crown and maximum foliage on maturity of the plant after three years, when it starts yielding economically.

Plant Protection

E. marcorhyncha is very susceptible to termite damage which increases when the plant loses vigour. To eliminate such attacks, doses of Chlordane or Heptachlor are incorporated in the soil repeatedly in the nursery stage. Such applications are sometimes needed after transplanting as well.

The leaf webber also attacks the plants during the monsoon and this can be easily controlled by spraying 0.1 % Metacid during the pre and post-harvest stage.

Harvesting and Yield

The growing trees yield the first harvest of foliage during the second year after the monsoon period and two harvests during the third year, before and after the monsoon. During the three years of plant growth and development, the plants are trained to a moderate height and dense crowns are developed. From the fourth year onwards, the plants successively yield three harvests of foliage in May-August and November-December, depending upon the weather conditions. The plantation produces economic yields for 10 years after three years of initial growth and development. There are records of 93-year-old plantations yielding rutin from their foliage in Australia, but with reduced rutin content.

A growing tree on an average yields 8 kg of foliage per tree from the third year onwards. This amounts to 10 to 12.5 t/ha of herbage.

Post-Harvest Handling of Leaves and Storage

The freshly harvested green leaves should not be stored for more than 3–4 days, because of possible decomposition and attack by insect pests and diseases. The leaf material should be dried, spread out in thin layers on clean yards for a couple of days, until the leaves are completely dried. The leaves may also be dried safely at temperatures of 80–100°C in artificial driers, with adequate air-circulation systems. The dried leaves are then packed in polythene bags for longer periods of storage.

Yield of Rutin

The rutin content varies from leaf to leaf in a plant and also lengthwise within a leaf. The content is maximum (20%) in young leaves and tips and decreases gradually to 3% in the mature leaves. The July-August harvest of 3-year-old plants yields the maximum rutin content, when compared to the other months of the year. Furthermore, the rutin content decreases slightly with the aging of the plant. Nevertheless, the average rutin content on moisture-free leaves is 5%. The moisture content in the freshly harvested green leaves range between 51 and 58%.

Seena

Plant Profile

Family	:	Leguminosae
English name	:	Senna
Indian name	:	Svarnapatri (Sanskrit) Sanay Sana ka Patt (Hindi) Nilavirai, Nelavakai (Tamil)
Species and varieties	:	Cassia angustijolia Vahl. C. acutifolia Del

		C. obovata
		C. italica
		C. abovata
		ALFT-2, Sona
Distribution	:	India, Southern Arabia, Yemen
Uses	:	Drugs

Senna (Cassia angustijolia Vahl.) is a small perennial undershrub, a native of Yemen, South Arabia. The leaves and """" pods contain Sennosides used for their laxative properties. It was introduced into Tamil Nadu in the eighteenth century where it is grown as an annual crop of 5 to 7 months duration in 8000 to 10000 ha both under rainfed and irrigated conditions. Successful cultivations have also been demonstrated in Karnataka (Bangalore) Gujarat (Anand), Rajasthan (Jodhpur) and Delhi with comparable yield and quality of the produce, but it is yet to establish itself as a cash crop there due to the lack of organized marketing. A major part of the produce is exported in the form of leaves, pods and sennoside concetrates, though several pharamaceutical houses utilize it for the manufacture of calcium sennoside granules, tablets and syrups for marketing within the country as well. Germany, Hungary, Japan, Netherlands and the USA are the main markets. The export of leaves and pods from India is of the value of Rs. 20 million annually. In addition, the erstwhile USSR contries buy sennosides (concentrates) and the current Cl'1nual export of the sennoside concentrate is around Rs. 20 million.

Senna is used in medicine as a cathartic. It is especially useful in habitual constipation. It increases the peristaltic movement of the colon. The pods have the tendency to gripe caused by senna may be obtained by combining it with an aromatic or a saline laxative. The laxative principles of senna are two glycosides, *viz.*, sennoside A and sennoside B. The plant also contains sennosides C & D. A number of other glycocides have also been reported. It also contains beta-sterol (0.33%) and flavanols-kaemferol, kaempferin and isorhamnetin. They are used in the form of decoctions, powders, confections and many other household preparations. Senna extracts, on storage, lose biological activity more rapidly than is revealed by chemical estimation. The leaves/pods as such or in powder form do not lose potency easily. The quantity administered in the form of infusion ranges from 0.6 to 2 g of leaves or pods soaked in 125 ml of warm water for about 12 hours, whereas, it is usually 30 mg of sennosides per tablet.

Origin and Distribution

Cassia is a pantropic genus comprising of 500 species, of which 20 are found in the native flora of India. There are two sources of the Senna drug, namely, Cassia augustifolia Vahl, and C. acutifolia. Del. C. angustifolia, commonly called Tinnavelly senna, is indigenous to Somalia, Southern Arabia, part of Sindh (Pakistan) and the Kutch area of Gujarat. In India, it is cultivated in Gujarat, Rajasthan and Tamil Nadu. In Tamil Nadu it is cultivated in the districts of Tirunelveli, Ramanathapuram, Madurai, Salem and Tiruchirapalli in an area of about 10000 hectares. C. acutifolia known as Alexandrian senna is indigenous to Sudan and Sinai. It is cultivated on a small scale in Sudan and Egypt, and is mainly collected from wild plants of C. acutifolia, whereas the Indian produce comes entirely from the cultivated crop of C. angustifolia and their pods contain about 2.4–4.5 and 1.25 to 2.5% sennosides, respectively.

Description of the Plant

Cassia angustifolia (2n = 28), belonging to the family Leguminosae, is a small perennial undershrub, below 1 m in height with ascending branches. The leaves are large, compound and pinnate. The full-grown leaflets are bluish-green to pale-green in colour and emit a characteristic fetid smell when crushed. The flowers are bright yellow in colour, arranged in axillary, (on subterminal) erect, many-flowered racemes. The flowers are not season-bound and are borne between 35 to 70 days of age, depending upon the time of sowing, and its season of cultivation. The pods appear immediately after flowering, are slightly curved, 3.5 to 6.5 cm long and up to 1.5 cm broad, green in the beginning changing to greenish-brown to dark brown on maturity and drying. Each pod has 5 to 7 ovate, compressed, smooth, dark-brown seeds.

The other species is very similar to C. angustifolia in morphology and growth, except in the dimension of its leaflets and pods. The leaflets are shorter and narrow. The pods are distinctly short, flat and broad, oblong but slightly curved upwards and are broadly rounded at the two ends. Further, surface making on the testa (seed) and stomatal index (leaves) are, remarkably, considered to distinguish between the two species. It has been found that Cassia obovata and C. italica also possess a fair to good percentage of anthraquinone compounds; C. obovata is common all over India. C. italica grows on sandy, saline soils in west Rajasthan and Saurashtra (Gujarat) and is a potential new source of the senna drug.

Varieties

The Gujarat Agriculture University, under the All India Co-ordinated project on Medicinal and Aromatic Plants at Anand, has released a late flowering type as 'ALFT-2' through selection, which remains in a vegetative stage till 100 days and is suitable for growing exclusively as a leaf-crop. The CIMAP, Lucknow, has also released a high-yielding variety named 'Sona'.

Soil

The crop can thrive on a variety of soils, but is largely grown on red loams, alluvial loams and the rich clayey rice-fields. The texture of the soils which account for the major hectarage under senna crop varies from sandy loam to loam, while the black cotton soils are heavier and more fertile. In general, these soils are more friable but less moisture retentive and of low fertility, usually, nitrogen and phosphoric acid are deficient throughout these soils. The average pH ranges from 7–8.5. It is very sensitive to water-logging and, hence, it is grown only on well drained soils.

Climate

Senna is a sun-loving crop and requires bright sunshine for its successful growth. The trials at Bangalore have shown that it can be grown as an early summer (Feb-March) or a winter (Oct-Nov) crop. However, sowing in the rainy season is not good, as the crop on harvest cannot be dried easily. Whereas, under North Indian conditions like Delhi and Gujarat, where the rainy season is short, it is reported to be the ideal time as the plants put on luxuriant growth. Heavy rains and cloudy weather during growth are harmful to the crop. An average rainfall of 25–40 cm, distributed from June to October is sufficient to yield good harvests.

Land Preparation

The land is prepared deep and exposed to the sun for 110–115 days to dry out the roots of perennial weeds, followed by two cross-ploughings, harrowing and levelling. FYM is incorporated into the soil at the time of the final cross-ploughing. Then the land is laid out into plots of convenient sizes with irrigation channels.

Cultivation

Propagation

The crop is raised from seeds. Since the seeds have a hard and tough seed-coat, a certain amount of abrading of its surface is necessary to induce quick germination. This is achieved by pounding the seeds lightly with coarse sand in a mortar.

Soaking the seeds for 10–12 hours before sowing is reported not only to give 100% germination, but also the stand of the plants grown is even. About 20 kg of seeds are required to cover a hectare of land.

Sowing

The seeds are broadcast or, more preferably, drilled at a distance of 30 cm in lines made at 30 cm apart at 1.5 to 2.5 cm depth on well-prepared land. The germination commences on the third day and is completed within a fortnight. Before sowing, the field should be perfectly levelled, lest it hamper uniform seed germination. It is found that treating the seeds with Thiram, Captan or Agroson G.N. at 2.5 g/kg protects the seedlings from damping-off and seedling blight diseases which are common occurrences.

The seeds can also be dibbled on the inner sides of the ridges opened at 45 cm distance, maintaining a plant-to-plant spacing of 30 cm. Only about 5 kg of seeds are required for this method of sowing.

Manures and Fertilizers

Farmers in Tamil Nadu use 4–5 cartloads (5–10 t/ha) of well-rotted FYM at the time of sowing. In most cases, it fully utilizes any leftover nutrient from the previous crop and, therefore, the nutrient requirement of senna is variable. However, trials in India have shown that the crop in all, takes 50–100 kg of N, 20–50 kg of P_2O_5 and about

30 kg of K_2O/ha in a growing period of 130–150 days, depending upon the growth and number of pickings. In general, where specific soil nutrient status of the field is not readily found, 80 kg of N and 40 kg each of P_2O_5 and K_2O/ha may be applied for the optimum production of leaf, pod and total alkaloids. Of these, the entire dose of P and K and 50% of N should be applied at the time of sowing, and the remaining 50% of N is to be applied 90 days after sowing.

Crop Rotation and Intercropping

In Tamil Nadu, senna is grown after the paddy and intercrop cultivation in between rows of cotton, sesamum, chillies, brinjal, okra and tomatoes is popular. Studies conducted at Delhi showed that sennamustard and senna-coriander rotations gave higher profits and these rotations have proved superior.

Irrigation

Senna can be economically grown under rainfed conditions. In most years, the crop needs no irrigations except during prolonged drought. However, when it is grown as a semi-irrigated crop, the yield increases considerably. In all, about 5–8 light irrigations are enough to raise a good crop of senna as heavy irrigations are injurious to the crop.

Weeding and Interculture

When the plants begin to grow, once or twice interculturing is given after which the rows close up. The first weeding-cum-hoeing is done at 25–30 days, a second at 75-80 days and a third at 110 days, to keep the soil free from weeds. The use of Teeflan herbicide as a pre-emergent spray at the rate of 4 kg/ha has been reported not only to increase the yield, but also the anthraquinone content.

Pests and Diseases

Pests

The larvae of several leaf-eating caterpillars feed on the green senna leaves. Usually, the spraying of Carbaryl (4 g/l) periodically in the growing season controls the infestation. The white butterfly (Catopsilia pyranthae) attack on this crop can be minimized by sowing the crop in March-April, instead of June-July in North India. Another pest, a pod borer is also reported to attack the pods and can be controlled by spraying Endosulphan (0.05%) or Carbaryl (0.25%) at an interval of 10–15 days.

Diseases

The leaf-spot disease is the most serious disease and causes severe damage to the crop. The infested leaves drop off, resulting in poor yield. The disease is caused by Alternaria alternata (Fr.) Keisel., Phyllosticta spp. and Cercospora spp. The sennoside content in the leaves is inversely proportional to the intensity of the disease. The spraying of 0.15% Dithane M-45 at fortnightly intervals, 3 times in a period of 5–6 weeks, has been found to control the disease effectively.

Damping-off of seedlings is the most devastating disease, which is caused by Rhizoctonia bataticola. The disease spreads rapidly if the fields are affected by stagnating moisture at this stage. Treatment with Thiram or Captan at 2.5 g/kg of seeds protects the growing seedlings. At a later stage its other physiological form, called Macrophomina phaseoli develops; it causes dry rot in the crop and kills the plant. The fields can be drenched with 0.2% Brassicol or 0.5 to 01.%. Rhizoctol, but this only gives partial control.

The leaf-blight disease is caused by Phyllosticta spp. usually occurring at the later stages of growth, in September-October. Cloudy days and humid weather conditions favour the spread of the disease. The leaf-spot disease initially appears as small, brown, irregularly scattered lesions on leaves which grow bigger in size and turn black in colour at a later stage. The spraying of 0.15% Dithane M-45 at forthrightly intervals can control the disease.

Harvesting, Processing and Storage

It has been found that young senna leaves and pods contain a high sennoside content but since the produce is sold on the basis of weight, a balance between weight and content has to be made, to choose the right stage for harvest. It is also found that senna plants produce foliage containing higher sennosides between 50–90 days of sowing. The first picking of the foliage crop should be done at 50–70 days' age, depending upon the total plant growth. The picking of leaves is done by hand so that most of the growing tops are removed at harvest; this also induces more branching which, otherwise, reduces the foliage growth considerably. A second picking is taken at 90–100 days and the third picking between 130–150 days, when the entire plants are removed, so that the harvested material includes both leaves and pods together.

The harvested crop should be spread in a thin layer in an open field to reduce its moisture. Further drying of the produce is done in well-ventilated drying sheds. It takes 10–12 days to dry completely. The dry leaves and pods should have a light-green to greenish-yellow colour. Improper and delayed drying changes the colour to black or brown which lowers the sennoside content and it fetches a lower price. A rapid mechanical drying at 40°C could also be attempted. The seeds contain no sennosides and only add weight to the produce. The pods are threshed during drying to remove the seeds. The produce is baled under hydraulic pressure and wrapped in gunny bags for export.

Yield

A good crop of senna can give 15 q/ha of dry leaves and 7 q/ha of pods, under irrigated and good management conditions. The yield, under rainfed conditions, is about 10 q/ha of leaves and 4 q/ha of pods.

Estimation of Sennosides

To 0.5 g (40 mesh) of accurately weighed powder in a 100 ml flask add 50 ml of water, swirl to mix, attach a reflux condenser and immerse the flask in a boiling water-bath with the water-level in the bath just above that of the liquid in the flask, for 15 minutes. Allow the mixture to cool. Decant and filter the extract into a volumetric flask (100 ml). Add 50 ml water to the residue and extract again in the boiling water-bath as above. The final volume in the flask is made up to 100 ml.

Transfer 20 ml of the clear, supernatant liquid from the volumetric flask to a separating funnel, add 0.05 ml of hydrochloric acid and extract with two successive 15 ml portions of chloroform. Discard the chloroform extracts. Transfer 20 ml of aqueous extract from the separating funnel to a 100 ml flask. Adjust the pH to 7-8 (with pH paper) with 5% w/v solution of sodium carbonate in water. Add 20 ml of 6.3% w/v freshly-prepared solution of ferric chloride($FeCl_3$) in a water-mix, attach a reflux condensor to the flask and heat on a boiling water-bath for 20 minutes and without cooling the liquid, add 2 ml of HCl. Continue the heating for a further 20 minutes, swirling the liquid in the flask frequently, to dissolve the precipitate and allow it to cool.

Transfer the liquid into a separating funnel, extract with three 25 ml portions of solvent ether, having used each portion to rinse the flask first. Wash the combined etheral extracts with two 15 ml portions of water and dilute to 100 ml with solvent ether.

Evaporate 10 ml (exact) of the etheral solution to just-dryness on a water-bath and dissolve the residue in 10 ml of 1N KOH. Then, immediately measure the extinction of 1 cm layer to 500 mm using water as a blank.

Steroid-bearing Solanum

PLANT PROFILE

Family	:	*Solanaceae*
English name	:	Steroid-bearing Solanum
Indian name	:	Kasi Kateri (Hindi), Kasibadane (Kannada)
Species and varieties	:	*Solanum viarum* Dunal. Arka Mahima, Arka Sanjeevini BARC Strain, Galxo strain Pusa-l, RRL 20-2, RRL-GL-6

Distribution : India, Burma, China
Uses : Drugs

Among the various plants which are being used as raw materials for the production of steroidal drugs, steroid-bearing olanum (Solanum viarum Ounal.) holds an important place due to its quick growth and low initial investment in its commercial cultivation. It yields a glyco-alkaloid, solasodine, a nitrogen analogue of diosgenine. Solasodine through 16-dehydro-pregnenolone (16 DPA) is converted to a group of compounds like testosterone and methyl-testosterone and corticosteroids like predinisolone and hydrocortisone. These steroidal compounds have anti-inflammatory, anabolic and antifertility properties, due to which they find large-scale use in health and family planning programmes all over the world.

Distribution

It is widely distributed in the subcontinent, extending from sea-level up to 2 000 m and is reported from Khasi, Jaintia and the Naga Hills of Assam and Manipur. It occurs in Sikkim, West Bengal, Orissa, the Upper Gangetic Plains and in the Nilgiris, ascending to an altitude of 1600 m. It is reported from North-east, North-west, southern as well as Central India, and extends into Burma and China. Its commercial cultivation is mainly confined to the Akola-Jalgoan tract of Maharashtra in an area of about 3000 ha.

Description of the Plant

Solanum viarum (2n = 24), belonging to the family Solanaceae, is a stout, branched, woody shrub attaining a height of 0.75 to 1.5 m. The stem has spines, the leaves are ovate to lobed with spines on both the surfaces, the flowers are hermophrodite, borne on axillary clusters, white; the berries are yellowish when ripe or greenish; the seeds are small, brown in colour and abundant, embedded in a sticky mucilage.

Varieties

Some of the promising, less spiny strains developed in this crop are the Galxo strain, BARC Strain, Pusa-1, RRL 20–2 and RRL-GL-6, which is said to be completely spineless. 'Arka Sanjeevini' and

'Arka Mahima' are the two diploid and tetrapoloid varieties of this crop developed at the Indian Institute of Horticultural Research, Hessarghatta, Bangalore and released for cultivation.

Soil

Solanum viarum is a hardy plant and can be cultivated on a wide range of soils under various agroclimatic conditions, but it cannot withstand water-logging. Though it can be grown in all types of soils the best soil for its successful cultivation is red lateritic soil with a moderate quantity of organic matter. The plant does not perform well in very clayey soils.

Climate

It is found growing under different kinds of climates throughout the length and breadth of the country, but prefers a moderate climate for its successful growth. The growth as well as development of the plants and, finally, the yield of fruits depend significantly on the climatic conditions. Depending upon the temperature, dry period length, rainfall and irrigation facilities, the sowing time may vary from June to September or October. The plants sown in late October to February are more susceptible to virus diseases, and sowing in March is not practicable as harvesting and drying will be hindered by the rainy season. Under rainfed conditions, the crop should be sown from mid-June to the first week of July. Under irrigated conditions the sowing is September results in profuse flowering and berry formation during the winter months.

Land Preparation and Planting

Well before the time of transplanting, the field is thoroughly prepared, about 25 t/ha FYM is incorporated into the soil, it is divided into blocks of convenient sizes, and ridges are opened at a distance of 60 cm.

At the time of transplanting, the seedlings uprooted from the beds are thoroughly cleaned of soil and dipped in 0.1 % Bavistin solution for at least 5 minutes to overcome the wilt problem and, thereafter, transplanted at a distance of 45 cm within the row. Spacings of 50 × 50 cm, 75 × 75 cm and 90 × 120 cm are also recommended for this crop for different locations.

Cultivation

Nursery Raising and Transplanting

Although the crop can be raised by broadcasting the seeds directly into the field, it is advantageous to raise a nursery. The seedlings are raised in nursery-beds of 10 m × 1 m size which are convenient for weeding, irrigation and removing seedlings. To each strip, 5 baskets of FYM and 1 kg of calcium ammonium nitrate are applied. Seeds which are presoaked in water for 24 hours are sown in lines of 1 to 1.5 cm-deep furrows, 10 cm apart and covered with a thin layer of soil. About 1.25 kg seeds sown in 5 nursery beds of the above size will provide enough seedlings for planting 1 ha of land. Their germination is completed in 7–10 days. Within 4-5 weeks, when the seedlings are 10–12 cm high and develop 6 leaves, they are ready for transplanting into the main field.

Manures and Fertilizers

Solanum viarum responds well to the application of manures and fertilizers. Green manuring before planting has been found to increase the yield by 20%. For better yields, an application of 100:60: 40 kg of N, P_2O_5 and K_2O/ha, depending upon the soil condition, is recommended. The entire quantity of phosphorus, photassium and half of the nitrogen are applied at the time of land preparation. While the remaining half of the nitrogen is applied when the plants start flowering. A fertilizer dose of 65: 40: 40 kg of N, P_2O_5 and K_2O has been recommended under Banglaore conditions.

Irrigation

In the absence of sufficient moisture in the soil, the field should be immediately irrigated after transplanting. The crop is further irrigated at weekly intervals during the first month and then the interval is increased to once in fortnight, and later as and when required.

Interculture

After 2-3 weeks of transplanting the first weeding is done, and later when the crop is 2–3 months old. Afterwards, the crop puts on enough canopy to smother the surface and, hence, no growth of weeds takes place.

Pests and Diseases

Insect Pests

The plant is sufficiently hardy and, therefore, free from any of the serious pests. However, it is sometimes attacked by leaf-eating caterpillars and wingless hoppers. In case of severe attack, the crop may be sprayed with Endosulphan to control them. The fruit borer (Leucinodes orbonalis) and root-eating grubs are the other pests reported. Ekalux (2 ml/l) can be sprayed to control the fruit borer. Chloropyriphos may used to control grubs.

Diseases

Some of the diseases the crop is attacked by are: powdery mildew, collar rot, bacterial blight, leaf-blight and damping-off.

Powdery mildew will be noticed during prolonged dry and warm period, Bavistin (1 g/l) may be sprayed to control this disease.

Collar rot or Fusarium wilt can be overcome by keeping the field clean and planting the crop in a well drained soil. Dipping the roots of the seedlings in a 0.1 % solution of Bavistin for 1 hour and drenching the seed-beds with 0.25% of copper oxychloride or 0.1 % of Bavistin solution can control the disease. Sometimes the plants are attacked by mosaic, caused by three different viruses, which leads to stunted growth and chloratic leaves. Such plants are better removed and destroyed. Bacterial blight is not a serious disease on this crop. However, under severe incidence, a treatment with the solution of 30 g of Streptocyclene and 30 g of Copper sulphate dissolved in 500 1 of water per hectare controls this disease.

Leaf-blight is caused by Pythium butleri and is not a serious problem in this crop.

Seedlings may also be attacked by damping-off disease while growing in the nursery-beds. To control this, the seeds are treated with Bavistin @ 3g /kg of seeds. Drenching the seed-beds 2-3 days after sowing and again when 50% of the seeds have germinated, with 0.1 % of Bavistin or 0.25% of copper oxychloride solution, helps to check the further spread of the disease.

The occurrence of the chlorotic stunt disease and its association with the root-knot nematode (Meloidogyne jauanica), and the wilting of plants due to M. incognita have also been reported.

Harvesting and Processing of the Berries

Generally, the accumulation of glyco-alkaloid increases with the physiological age of the fruit and attains its peak value in the fruits of 50-60 days. This state of fruit growth coincides with the change in fruit colour from green to just yellow with streaks of green still present, after which the glyco-alkaloid content falls gradually with the maturity of fruits. The alkaloid is distributed throughout the fruit. It is, however, established that about 60% of this is present in the seeds and the remaining 40% in the pericarop.

The crop takes about 6 months to be ready for harvesting. Harvesting is one of the labour-intensive operations in the cultivation of S. uiarum. The spiny nature of the plant hampers plucking the berries at the right stage of maturity, which is very important. During the' first part of the harvesting season, when the fruit is big, on an average, one person with gloves can pluck about 50 kg of berries, while working 8 hors a day. However, some good workers in the peak season when most of the berries are ready for harvest can pick even 80 kg of berries per day. The figure is reduced to 40 kg towards the closing when the fruits become smaller in size. The picking operation reads over 3 months, because the fruits mature at different times. The processing of berries for marketing requires a iot of care. Fresh fruits contain about 80% moisture. The pharmaceutical firms need berries containing about 10% moisture. Hence, the berries must be dried in the sun.

Yield

When the crop is grown by adopting proper cultivation practices, it may yield nearly 10000 kg/ha of fresh berries which, in turn, will give about 2500 kg/ha of dried berries.

Sweet Flag

PLANT PROFILE

Family	:	Aeraceae
English name	:	Sweet Flag
Indian name	:	Ugragandha (Sanskrit), Bacc, Gorbacc (Hindi) Baje (Kannada), Vashampu (Tamil)
Species and varieties	:	Acarus calamus L. A. gramineus Soland
Distribution	:	India, North America, Asia, CentraL Europe Japan, Rumania, USSR
Uses	:	Aromatics, Confectionery, Drugs Flavouring Liqours, beer, gin, vinegar snuff and various other preparations, substitute for Ginger, Insecticide

Sweet flag (Acarus calamus L.) is a herb belonging to the family Aeraceae. The dried root (rhizome) has long been employed in medicinal preparations and for flavouring liquors. It contains a volatile, yellowish-brown oil of peculiar, but pleasant, slightly sweet odour, that can be isolated by steam distillation. The yield and physico-chemical constituents and, consequently, the composition of the 'calamus oil' depend upon the source from which the rhizomes are obtained. The European and American oils are similar in composition, consisting of low amounts of 'asarone' which can be compared to the Indian oil. The rhizomes contain a high percentage of starch.

The plant is aromatic on account of the presence of an essential oil, which contains the glucosidic bitter principle, 'Acorin'. It has an expectorant action due to the presence of the essential oil and is commonly used as a remedy for asthma. It is also used as a remedy for chronic diarrhoea. The rhizome is an appetizer and is also useful in the treatment of epilepsy, delirium, hysteria and loss of memory. In Unani literature, the rhizome has been described as having a very bitter taste and is said to be useful as a brain tonic. The powdered rhizomes have been employed in combination with Rauwolfia serpentina and other drugs in the treatment of neurosis, insomnia and hysteria. Because of the pleasant aromatic smell, the rhizomes are extensively used as an aromatic agent in agarbathies, dhoops and havana samagries. The powdered rhizome and the essential oil can be used as a safe insecticide against flies, mosquitoes, bed-bugs, moths and lice. This property is due to the presence of the 'trans-isomer of asarone'. The leaves and rhiozmes are used to control lice. It's essential oil is used in flavouring liquors, beer, gin, vinegar, snuff and various other preparations. The "bach" of commerce, which has many medicinal uses is prepared from the rhizomes. The fresh rhizomes are used in confectionery and also as a substitute for ginger. In moderate doses, the oil produces antispasmodial action on the involuntary muscle tissue, inhibiting the excessive peristaltic movements of the intestine.

Orgin and Distribution

It grows wild on the edges of swamps, on the banks of rivers and ponds in North America, Europe and Asia. The countries growing this crop commercially are the USSR, Central Europe, Rumania, India and Japan. In India, the leading states are Karnataka, Kashmir, Manipur, Arunachal Pradesh, Meghalaya and in the foothills of the Himalayas, up to 1800 m. In Karnataka, it is being grown in the Koratagere Taluk of Tumkur district in an area of 100 acres and the annual production is 400 t of dried rhizomes.

Description of the Plant

Sweet flag is a monocot plant. The plant is a herb, with narrow, 0.75–1.5 m, long leaves. The breadth of the leaf is 2–4 cm and resembles the leaves of an iris. The leaves are bright green, thickened in the middle and the margin is wavy. The spathe is 6–30 cm long and the spadix is slightly curved. The fruit is a berry, yellow-green in colour, angular and 1–3 seeded.

The economic part of the plant is the underground rhizome. The rhizomes are horizontal and jointed, with a spongy texture. The rhizomes are 1.5–2.5 cm thick and sometimes up to 2 m long.

Species/Varieties

Acorus is a small genus of herbs, comprising 2 species.

Acorus Gramineus Soland

It is a Japanese species, occasionally encountered in Sikkim (up to 1800 m) and the Khasi hills (1200–1 400 m). It is used for the same medicinal purpose as A. calamus. Here, the mid-rib of the leaf is absent. It is a semi-aquatic, perennial herb with creeping and many-branched aromatic rhizomes of spongy texture, from 1.5 to 2.5 cm thickness and sometimes several metres long. The leaves are 10–22 cm in width and 0.60–1.8 m in length. The rhizome will be brown outside and white and sponge-lined inside.

Acorus Calamus L.

It is a semi-aquatic perennial herb with a stout midrib, with a creeping and many branched aromatic rhizome. It is found growing wild or cultivated throughout India and Sri Lanka ascending to 1800 m in the Himalayas. The plant thrives in marshy places and moist situations, like the edges of lakes and banks of streams and rivers. It is plentiful in the marshy tracts of Kashmir and Sirmoor in Manipur and the Naga Hills. It is also grown in North America, Europe and Asia.

Soil

Sweet flag is cultivated in almost the same way as rice and may be grown in any part of the country where suitable irrigation facilities are available. Although calamus in the wild state grows in water, it may be cultivated in almost any good, but fairly moist soil. It is usually grown in areas where paddy can be grown. It is generally grown in the clayey loam and light alluvial soils of river banks.

Climate

Sweet flag is a hardy plant and is thus found growing from tropical to subtropical climates. However, it prefers an area with good and well-distributed rainfall throughout the year.

Planting Season

The best planting season is spring (March-April), but the crop may be planted almost at any time of the year taking into consideration the fact that plenty of sunshine should be available at harvest time to dry the rhizomes.

Planting Seasons

The field is laid out exactly as for rice, irrigated sufficiently and, after ploughing twice, is watered heavily and ploughed again. The field is then left alone for a few days and puddled. After that, it is levelled with a levelling board.

Bibliography

Akhtar Hussain "*Belladonna and its cultivation in India*", Farm Bull., 5, Central Institute of Medicinal and Aromatic Plants (CIMAP), Lucknow, 1977

Akhtar Hussain "*Pyrethrum and its Cultivation in India*", Farm Bull., 4, CIMAP, Lucknow, 1977

Akhtar Hussain, *Medicinal Plants and their Cultivation*, CIMAP, Lucknow. Anjana, 1993

Akhtar Hussain, O.P. Virmani, Ashok Sharma, Anupkumar, L.N. Mishra, *Major Essential Oil Bearing Plants of India*, CIMAP, Lucknow, 1988

Badhwar, RL., P.S. Rao and H. Sethi, *Some Useful Aromatic Plants, Forest Research* Institute (FRI), Dehra Dun, 1964

Bebusova, L.S. and L.v. Denisova, *Rare Plants of the World*, Oxford and IBH Publishing Company, New Delhi, 1991

Bradu, B.L., and C.K Atal, "Cultivation of *Ammi majus* L. in Jammu", J. Pharm., 32: 145-147, 1970

Caius, J.E, *The Medicinal and Poisonous Plants of India*, Scientific Publishers, Jodhpur, 1986

Chakravarthy, D.K, Khatva and T.K Das, "*Pathological problems in the cultivation of Vinca rosea*", Indian Agri., 20 (4): 284-295, 1976

Chandra, v., P.N. Mishra and A Singh, *Matricaria for Blue Oil*, National Botanical Research Institute (NBRI), Lucknow, 1979

Chopra, R.N., Chopra, I.C. and Verma, B.S. Supplement to Glossary of *Indian Medicinal Plants*, CSIR, New Delhi, 1969

Chopra, R.N., Nayar, S.L. and Chopra, I.C. Glossary of *Medicinal Plants of India*, CSIR, New Delhi, 1965

Cone, N.J., R Milner and M. Neuse, Analysis of Vinca Alkaloids by TLC, J. Pharma. Sci., 52 (7): 688-92, 1963

Cordell, G.A. "*Anticancer agents from plants*", Progress in Phytochemistry, 5: 273-316, 1977

Dalal, KC., M.A. Patel, "Guggal", in *Advances in Horticulture*, Eds: KL. Chadha and Rajendra Gupta, 11: 491-502, 1995

Dastur, J.E. *Medicinal Plants of India and Pakistan*, D.B. Taraporevala Sons and Co. Ltd., Bombay, 1952

Devi, S., Yield and Quality Constituents in Costus speciosus under Varying Levels of N, P and K, M.Sc. (Hort) Thesis submitted to the Kerala Agricultural University, Thrissur, 1981

Elgamel, S.A., EA El-Omar and M.M. Mahmud, "*Effect of nitrogenous fertilizers on the growth and yield of Hibiscus sabdariffa L.*", Acta Hart., 144: 135-144, 1984

Elizabeth Peplow, The Herb Book, W.H. Allen & Co. PLC, London, 1984

Farooqi, AA and M.M. Khan, *Production Technology for Medicinal and Aromatic Crops*, University of Agricultural Sciences, Bangalore, 1993

Gopinath, G.T, M.R. Narayana and Akhtar Hussain, "*Duboisia and its cultivation*", Farm Bull. 19. CIMAp, Lucknow, 1994

Guenther, E. Essential Oils, Van Hostrand Co. Inc., New York, 1950

Gupta, R, V.K Srivastava and M.L. Maheshwari, "*Sennoside content in senna leaves as influenced by growth and nutrition*", Indian J. Pharm., 39 (5): 109-111, 1977

Gupta, R. "Senna has a growing export market", Indian Fmg., 21 (8): 29-32, 1971

Handa, S.S. and M.K Kaul, *Cultivation and Utilization of Medicinal Plants*, CSIR, New Delhi, 1982

Handa, S.S. and M.K. Kaul, *Supplement to Cultivation and Utilization of Aromatic Plants*, RRL, CSIR, Jammu, 1997

Hans Flick, Medicinal Plants, W. Foulsham and Company Limited, London, 1970

Hegde, D.M. "*Intercropping of medicinal yam with short duration cowpea, cluster bean and kidney bean*", Indian J. Agric. Sci., 51 (4): 262263, 1981

Jain, S.K. Dictionary of *Indian Folk Medicine and Ethnobotany*, Deep Publications, New Delhi, 1991

John Lindly, *Flora Medica: A Botanical Account*, Ajay Book Service, India, 1981

Joshi, B.S., S. Ramanujam and M.B.L. Saxena, "*Improvement of some essential oil bearing spice plants*", Information Bull., RRL, Jammu. 18: 94-100, 1963

Kafaga, E.R, D. Frinz and S. Rahm, "*Some physiological reactions affecting yield of Roselle* (Hibiscus sabdariffa L. var. Sabdariffa)", Temperlandwirat, 1980

Kanny Lall Dev Rai Bhadur, "*The Indigenous Drugs of India. The Chronica Botanica, Pama Primlane*, New Delhi, 1973

Kapoor L.D., P.K Dutta and Ajit Singh, "*Cultivation of Matricaria chamomilla*", Bull., RRL, Jammu, 1 (2): 104-106, 1963

Mamatha, H., A.A. Farooqi, S.S. Joshi and TG. Prasad, "Pollen studies in Gloriosa superba, Linn.", Crop Res. J., 5 (1): 92-97, 1992

Mamatha, H., AA Farooqi, S.S. Joshi and TG. Prasad, "*Studies on floral biology in Gloriosa superba Linn.*", Crop Res. J., 5 (1): 85-91, 1992

Meenaxi, S. "T*hey make money with senna in the South*", Jndian Fmg.,8: 5, 1957

Muddappa, *Medicinal Plant* Source Book, International Library Association, Dehra Dun, 1992

Nadakarni, P. *Indian Medicinal Plants*, Orient Longman Ltd, Hyderabad, India, 1992

Patil, S.L., Some Issues on *Agricultural Technology*, ed. K.R. Ganapathy, UAS, Misc. Series. 35, 1987

Rasheeduz Zafar, *Medicinal Plants of India*, CBS Publishers and Distributors, Delhi, 1994

Singh,V.P., "Cultivation of the Egyptian herbs *Ammi majus* and *Ammi visnaga*, Indian Forester, 32: 555, 1963

Thankur, R.S. Puri, H.S. and Akhtar Hussain, *Major Medicinal Plants of India* CIMP, Lucknow, 1986

Viswanathan,T.V., "Long Pepper", in *Advances in Horticulture*, eds. K.L. Chadha and Rajendra Gupta, 11: 373-383, 1995